TRIGONOMETRY
WITH
APPLICATIONS

M. N. Manougian

Department of Mathematics
University of South Florida

Mariner
Publishing
Company   Tampa, Florida

© Copyright 1980 by Mariner Publishing Company, Inc.
Tampa, Florida

The Author is grateful to the Dellen Publishing Company for
permission to use materials from *Basic Concepts of College
Algebra and Trigonometry*, by M.N. Manougian, © copyright
1978 by Dellen Publishing Company.

Printed in the United States of America

Cover Design and Art: Joe Traina and Jake Fernandez

Library of Congress Cataloging in Publication Data

Manougian, M   N
    Trigonometry with applications.
    Includes Index.
    1. Trigonometry.   I. Title.
QA531.M375   516.2'4   79-28391

ISBN 0-936166-00-2

PREFACE

This text is the result of extensive classroom experimentation and is intended for use in a first course in college trigonometry. The text contains the traditional subject matter with special emphasis being placed on informal presentation, motivation, relevance, and testing.

There are seven chapters and appendices. Each chapter is divided into sections which begin with a statement of objectives. Mathematical concepts are introduced with simple examples, many of which illustrate the applicability of trigonometry to real world situations. Numerous worked-out examples are given to illustrate the concepts presented, and techniques and drills are emphasized throughout the text. Each section is concluded with a large number of drill exercises. The student is made aware that it is by doing mathematics that one learns mathematics. The exercises which are graded in difficulty, are designed to help the student's understanding of the concepts discussed in each section. Following each chapter is a review emphasizing the main ideas discussed in the chapter. These serve as self-tests. Also, at the end of each chapter is a sample test.

In order to help students in their learning process, other features are also included. In each section, the student is asked to respond to simple questions under the heading "Self-test". The answers to these questions are provided at the bottom of the page for instant feedback. This feature is used to ensure student understanding of the material discussed and to prepare them for the exercises. Also, these serve as stopping points for the instructor who wishes to divide each section into subsections. Metric units are used extensively and a brief summary of Metric to English and English to Metric conversions are included in the Appendix. Finally, for better understanding and general interest, historical developments of some of the concepts and biographical sketchs of some of the famous mathematicians are presented.

The book is designed for a one-semester or a one-quarter course. Instructors may wish to create a syllabus and set their own pace and areas of emphasis. Various syllabi may be devised for a course. Chapter One is a review of functions which may be omitted for students with a course in college algebra. The trigonometric functions are introduced as functions whose domain is a subset of all angles. The circular functions are also introduced and the relationship between the two is indicated. Although computations are included, they are de-emphasized in view of the wide use of scientific calculators. In fact, this is why the chapter on logarithms is placed in the Appendix to be used as the need arises.

The author wishes to thank Janice Kartsatos for proofreading and checking the answers to the examples and exercises, the many students at the University of South Florida who have participated in the preparation of the manuscript, and the several prepublication reviewers for their comments and helpful suggestions.

M.N. Manougian
Tampa, Florida

TABLE OF CONTENTS

TRIGONOMETRY WITH APPLICATIONS

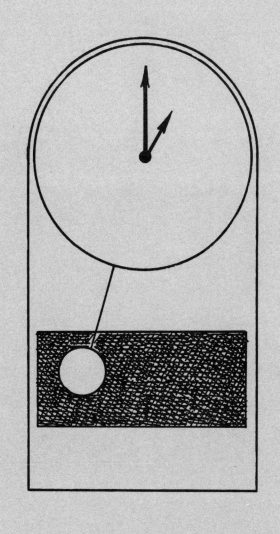

CHAPTER ONE:
REVIEW OF FUNCTIONS

1.1 THE RECTANGULAR COORDINATE SYSTEM

OBJECTIVES

1. *Define Cartesian product.*
2. *Introduce the concept of one-to-one correspondence.*
3. *Construct the number line.*
4. *Set up the Rectangular coordinate system.*

In this section we present a type of set* called the **Cartesian product**. This set is used to describe the association between two sets of numbers or objects. In everyday life we often encounter associations between two sets. For example,

interest earned and amount invested
time of day and temperature
month of year and the rise in food prices
students in this class and grades
cost and profit
distance traveled by a car and its speed.

RENE DESCARTES (1596-1650)

Descartes' coordinate geometry provided a vehicle whereby points in the plane could be viewed as ordered pairs of numbers and ordered pairs of numbers could be viewed as points in the plane. Thus, coordinate geometry unified algebra and geometry, enabling us to do geometry problems by algebra. The concept of the Cartesian product gives us an analytic way of viewing the geometric concept of dimension. Thus, one-dimensional space is represented by R, two-dimensional space by R² and so on.

*For a brief review of sets see Appendix A.1.

It might be helpful to take a moment and make up another list showing an association between two sets. The following two examples give some motivation for introducing the idea of a Cartesian product. This concept will then be used to introduce the Rectangular coordinate system.

EXAMPLE 1. Political analysts study the candidates of different political parties by pairing the leading candidate of one party with the leading candidate of another party. Suppose in a presidential election, the Republicans have three leading candidates, denoted by R_1, R_2, and R_3, and the Democrats have two leading candidates denoted by D_1, and D_2. All possible two-man races can be represented by the set of ordered pairs

$$\{(R_1,D_1),(R_2,D_1),(R_3,D_1),(R_1,D_2),(R_2,D_2),(R_3,D_2)\}$$

It might be helpful for the student to visualize the ordered pairs as the array shown in Table 1. In each ordered pair we have two entries, a Republican candidate and a Democratic candidate in that order.

	D_1	D_2
R_1	(R_1,D_1)	(R_1,D_2)
R_2	(R_2,D_1)	(R_2,D_2)
R_3	(R_3,D_1)	(R_3,D_2)

Table 1

	Cafe-c	Antonio's-a
ballet-b	(b,c)	(b,a)
jazz-j	(j,c)	(j,a)
play-p	(p,c)	(p,a)

Table 2

EXAMPLE 2. Michael and Terri are discussing their date for the evening. Michael suggests the following choices: a ballet performance, a jazz concert, or a play. Terri suggests that after the show they go to the Cafe de Paris restaurant or to Antonio's Pizza Parlor. Assuming that they are going to a show first and eat later, Michael and Terri can spend their evening in one of six ways. We illustrate the different possibilities in the array shown in Table 2.

Using set notation, we let $A = \{b,j,p\}$ and $B = \{c,a\}$. From these two sets, we form a new set called the Cartesian product of A and B and denoted by $A \times B$ (read "A cross B").

$$A \times B = \{(b,c),(j,c),(p,c),(b,a),(j,a),(p,a)\}$$

For instance, one of the entries of the set $A \times B$ is (j,a), which represents the possibility of going to the jazz concert first and then going to Antonio's Pizza Parlor. The set $A \times B$ consists of all possible pairs of elements in which the first element belongs to A and the second element belongs to B.

A pair of elements in which the order is specified is called an **ordered pair of elements**. The symbol (j,a) denotes an ordered pair of elements and the set $A \times B$ consists of ordered pairs of elements. The element j is called the **first component** of the element, and the element a is called the **second component** of the ordered pair (j,a).

In general, the **Cartesian product** (or the **cross product**) of two sets is defined as follows:

> **DEFINITION 1.1** If *A* and *B* are two nonempty sets, then the *Cartesian product* of *A* and *B*, denoted by $A \times B$, is the set of all ordered pairs (a,b) such that $a \in A$ and $b \in B$. That is

$$A \times B = \{(a,b) \mid a \in A \text{ and } b \in B\}$$

Since the order is important, the ordered pair (a,b) is generally different from the ordered pair (b,a). Thus, in Example 2, $A \times B \neq B \times A$. In fact, if Michael and Terri are starving, it makes a big difference whether they consider $A \times B$ or $B \times A$!

Self-test. If $A = \{0,1\}$ and $B = \{x,y\}$, then:

(a) $A \times B =$ (0,x) (0,y) (1,x) (1,y)

(b) $B \times A =$ (x,0) (x,1) (y,0) (y,1)

(c) $A \times A =$ (0,0) (0,1) (1,0) (1,1)

Now, consider the set of all real numbers *R*. The Cartesian product $R \times R$, which may be denoted by R^2, is the set of all possible ordered pairs of real numbers.

$$R \times R = \{(x,y) \mid x,y \in R\}$$

Before we discuss a pictorial representation of R^2 let us consider the number line. A convenient way to picture the real numbers is to represent them geometrically as the set of all points on a straight line. This will serve as an aid to understanding relations among real numbers. We shall assume the following:

For each real number, there corresponds one and only one point on the line, and conversely, for each point on the line, there corresponds one and only one real number.

This property between two sets, the set of real numbers and the set of points on a line, is called a **one-to-one correspondence**. For example, the two sets $A = \{a,b,c\}$ and $B = \{0,1,2\}$ can be put into a one-to-one correspondence, since for every element of *A* we can assign a unique element of *B* and vice versa. The sets $C = \{x,y,z\}$ and $D = \{0,1\}$ cannot be placed in a one-to-one correspondence. For example, we may assign x to 1 and z to 0, but there will be no element left in *D* that can be assigned to y in C.

ANSWERS:

(a) $A \times B = \{(0,x),(0,y),(1,x),(1,y)\}$
(c) $A \times A = \{(0,0),(0,1),(1,0),(1,1)\}$
(b) $B \times A = \{(x,0),(x,1),(y,0),(y,1)\}$

We now illustrate the correspondence between the set of real numbers and the set of points on a line. Consider a line L and select:

(a)   A point O called the **origin** and associate it with the number zero.
(b)   A convenient point to the right of zero to represent the number 1.
(c)   The **positive direction** as the direction traveled when going from zero to one. The opposite direction will be called the **negative direction** of the line when going from zero to the left.

The line segment from 0 to 1 is called the unit segment and its length is chosen as the unit length. The line L is then called a **number line** (also called a **directed line** or **coordinate line**). Ordinarily (but not necessarily), such a line is drawn horizontally, and the choice of the representation of the 1 to the right of 0 is arbitrary (see Figure 1).

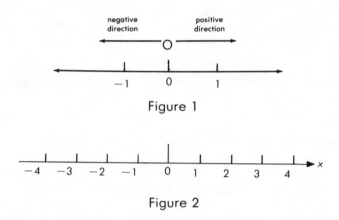

Figure 1

Figure 2

Next, we reproduce the unit length successively on both sides of 0 and 1 on the line L to obtain the graphical representation of the set of integers (see Figure 2). Note that the negative numbers are measured in the negative direction from 0.

Now we can construct a pictorial representation of R^2. We will use a **Rectangular** (or **Cartesian**) **coordinate system** to set up a correspondence between R^2 and the Cartesian plane, with each point in the plane corresponding to an ordered pair in R^2.

In order to set up a Rectangular coordinate system, we draw a horizontal line, which we call the **x-axis**. Then we draw a vertical line called the **y-axis** (see Figure 3). The point of intersection of the axes is called the **origin** and the two lines are called the **coordinate axes**. Usually the number zero is represented by the origin and a number scale is marked on each of the axes. On the x-axis, the positive numbers are to the right and on the y-axis, the positive numbers extend upward. The coordinate axes divide the plane into four **quadrants**. The four quadrants are:

Quadrant I: $\{(x,y) \mid x > 0,\ y > 0\}$
Quadrant II: $\{(x,y) \mid x < 0,\ y > 0\}$
Quadrant III: $\{(x,y) \mid x < 0,\ y < 0\}$
Quadrant IV: $\{(x,y) \mid x > 0,\ y < 0\}$

To avoid ambiguities, the coordinate axes are excluded from the quadrants.

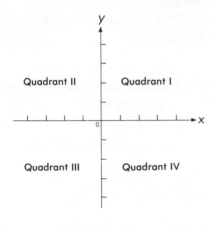

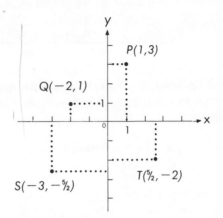

Figure 3   Figure 4

Looking at Figure 4, we see that the ordered pairs of numbers $(1,3)$, $(-2,1)$, $(-3, -\tfrac{5}{2})$ and $(\tfrac{5}{2}, -2)$ are represented by the points P, Q, S, and T in quadrants I, II, III, and IV, respectively. The ordered pair $(1,3)$ corresponds to the (unique) point one unit to the right of the y-axis (one unit in the positive x-direction) and three units above the x-axis (three units in the positive y-direction). The **coordinates** of the point P are the components 1 and 3 in the ordered pair of numbers $(1,3)$. The first component is also referred to as the **abscissa** and the second component is called the **ordinate**. The word coordinates was first used by Leibniz.*

---

* *GOTTFRIED WILHELM LEIBNIZ*

Gottfried Wilhelm Leibniz (1646-1716), philosopher and metaphysician was born in Leipzig, Germany, the son of a professor of philosophy. Leibniz is considered one of the greatest thinkers of modern times, proficient in law, theology, geology, and history, as well as serving as a diplomat. Leibniz wrote with ease in five different languages, Greek, French, Latin, English, and his native German. He received his bachelor's degree in philosophy at age seventeen and had a doctorate in law when he was twenty. During his lifetime, he invented one of the first calculating machines and founded the Academy of Sciences in Berlin.

In mathematics, Leibniz contributed immensely to the development of calculus and symbolic logic. In spite of his accomplishments, his life ended sadly, troubled in his final years by ill health and an ongoing bitter dispute with Newton over priority for the discovery of the calculus.

Self-test. Referring to the points in Figure 4, find:

    (a)   Abscissa of Q = _____

    (b)   Ordinate of Q = _____

    (c)   Abscissa of S = _____

    (d)   Ordinate of S = _____

It is important to note that to each element in R^2, there corresponds exactly one point in the plane, and to each point in the plane there corresponds exactly one element in R^2.

Now we consider some subsets of R^2.

EXAMPLE 3. Let $A=\{1,2,3\}$ and $B=\{4,5\}$. Find each of the following:

        (a) $A \times B$ (b) $B \times A$

Solution. We use Definition 1.1

    (a) $A \times B = \{(a,b) \mid a \in A \text{ and } b \in B\}$

               $= \{(1,4),(1,5),(2,4),(2,5),(3,4),(3,5)\}$

    (b) $B \times A = \{(b,a) \mid b \in B \text{ and } a \in A\}$

               $= \{(4,1),(4,2),(4,3),(5,1),(5,2),(5,3)\}$

In Figure 5, the set of circled points represents $A \times B$, while the set of points with squares represents $B \times A$.

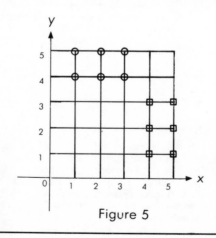

Figure 5

EXAMPLE 4. Let $A=\{t|-2\leq t<3\}$ and $B=\{2\}$. Find each of the following:

(a) $A\times B$ (b) $B\times A$

Solution. We use Definition 1.1

(a) $A\times B=\{(x,y)|-2\leq x<3 \text{ and } y=2\}$

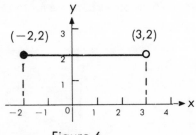

Figure 6

A geometric representation (or **graph**) of $A\times B$ is shown in Figure 6. Every element on the line between point $(-2,2)$ and $(3,2)$ represents an element of $A\times B$. The point $(-2,2)\in A\times B$. This is indicated by the shaded circle. On the other hand, $(3,2)\notin A\times B$, indicated by an unshaded circle.

(b) $B\times A=\{(x,y)|x=2 \text{ and } -2\leq y<3\}$

A geometric representation of $B\times A$ is shown in Figure 7.

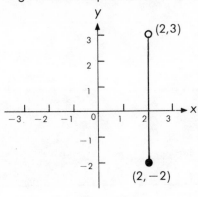

Figure 7

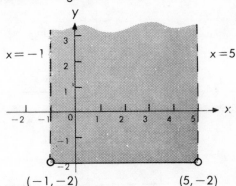

Figure 8

EXAMPLE 5. Let $A=\{t| |t-2|<3\}$ and $B=\{t|t\geq -2\}$. Find each of the following:

(a) $A\times B$ (b) $B\times A$

Solution. We use Definition 1.1

(a) $A\times B =\{(x,y)|x\in A \text{ and } y\in B\}$
$=\{(x,y)| |x-2|<3 \text{ and } y\geq -2\}$
$=\{(x,y)|-1<x<5 \text{ and } y\geq -2\}$

A geometric representation of $A\times B$ is shown in Figure 8. The shaded region, including the solid line joining the points $(-1,-2)$ and $(5,-2)$, represents elements that belong to $A\times B$. Note that points on the broken lines $x=-1$ and $x=5$ do not belong to $A\times B$.

(b) $B \times A = \{(x,y) \mid x \geq -2 \text{ and } |y-2| < 3\}$
        $= \{(x,y) \mid x \geq -2 \text{ and } -1 < y < 5\}$

A geometric representation of $B \times A$ is given in Figure 9.

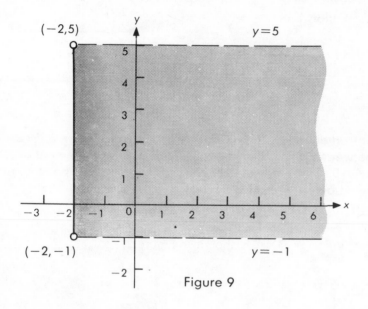

Figure 9

In Examples 3-5 above, the sets $A \times B$ and $B \times A$ are subsets of the set $R \times R$. Such sets are called **relations**. In general, a relation is any nonempty subset S of $R \times R$. We define the **domain** of the relation S to be the set of all first components of the elements of S and the **range** of the relation S to be the set of all second components of the elements of S. The geometric representation of the set S is called the **graph** of S. An important class of relations called functions will be studied in Section 1.3.

ANSWERS:

EXERCISES 1.1

In Exercises 1-10, locate each of the given points in a Rectangular coordinate system and indicate in which quadrant the point lies.

1. $(2,3)$ 2. $(1,-\frac{5}{2})$ 3. $(4,0)$
4. $(0,4)$ 5. $(-4,-2)$ 6. $(0,-3)$
7. $(-2,0)$ 8. $(-\frac{3}{2},\frac{3}{2})$ 9. $(6,-3)$
10. $(\frac{7}{2},-\frac{7}{2})$

In Exercises 11-18, specify the ordered pair representing the indicated points shown in Figure 10.

11. A   12. B   13. C
14. D   15. E   16. F
17. G   18. H

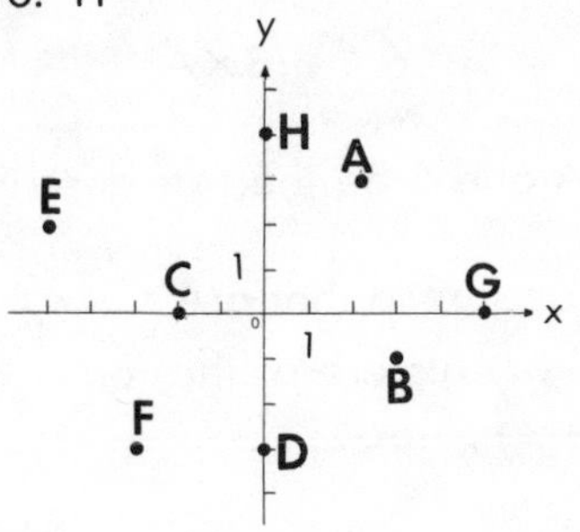

Figure 10

19. Jack and Liz are contemplating an outing. They decided to eat first, then go to a movie. If Jack suggested three restaurants and two shows to choose from, how many choices does Liz have? Represent the choices as a Cartesian product.

In Exercises 20-23, $A=\{x,y,z\}$ and $B=\{3,4\}$. Find:

20. $A \times B$ 21. $B \times A$ 22. $A \times A$ 23. $B \times B$

In Exercises 24-32, find the Cartesian products $A \times B$ and $B \times A$ and represent the resulting set geometrically.

24. $A=\{-1,1\}$, $B=\{2,3,4\}$
25. $A=\{1,2,3\}$, $B=\{1,2\}$
26. $A=\{t \mid -1 \le t \le 2\}$, $B=\{2\}$

27. $A=\{t\,|\,0<t\le 5\}$, $B=\{-1,2\}$
28. $A=\{t\,|\,|t-1|<4\}$, $B=\{-1,1\}$
29. $A=\{t\,|\,|t+2|<3\}$, $B=\{2\}$
30. $A=\{t\,|\,t\ge -1\}$, $B=\{t\,|\,|t+1|<2\}$
31. $A=\{t\,|\,|t|<2\}$, $B=\{t\,|\,t\le 4\}$
32. $A=\{t\,|\,|t-2|>3\}$, $B=\{t\,|\,t\le 1\}$

33. Use a map to find the coordinate of each of the following cities:
    (a) Los Angeles   (b) Philadelphia
    (c) New Orleans   (d) Columbia, SC

34. Let $A=\{1,2,3\}$ and let $B=\{1,2,3,4\}$. How many elements are in $A\times B$? $B\times A$?

In Exercises 35-37, let $A=\{1,2,3,...,k\}$ and let $B=\{1,2,3,...,m\}$ with $k<m$. Determine the given expression. [$n(A)=$number of elements in A].

35. $n(A\times B)$ 36. $n(B\times A)$ 37. $n(A\times A)$

In Exercises 38-40, let A and B be the sets described above.

38. Determine k and m so that $n(A\times B)=6$.
39. How many elements are of the form (x,x)?
40. Is $n(A\times B)=n(A)\times n(B)$?

1.2 THE DISTANCE FORMULA

One of the oldest results in mathematics is the famous **Pythagorean theorem**. Although it was discovered by the Babylonians, the general statement and proof are credited to Greek mathematicians. It was proved that in a right triangle, the area of the square constructed on the hypotenuse is equal to the sum of the areas of the squares constructed on the other two sides (see Figure 1). In terms of real numbers, if a and b are the lengths of two legs of a right triangle and c is the length of the hypotenuse, then

$$c^2 = a^2 + b^2$$

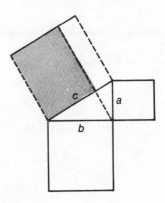

We shall use the Pythagorean theorem to establish an important formula. The student should recall the definition of absolute value and that $|x-y|$ is the distance between the two points corresponding to the numbers x and y. We shall extend the notion of distance to that of the distance between two points in the Rectangular coordinate system R^2. The distance between two points in R^2 is the length of the line segment joining the two points.

Figure 1

Let $A(x_1,0)$ and $B(x_2,0)$ be two points on the x-axis. The distance between A and B is

$$d(A,B) = |x_2 - x_1|$$

Now, consider the points $C(x_2,b)$ and $D(x_1,b)$, $b \neq 0$. The ordinates of C and D are equal and the figure ABCD is a rectangle, as seen in Figure 2. Since the lengths of opposite sides of a rectangle in a plane are equal, we assign the same length to the line segment CD as we did to line segment AB. Thus

$$d(C,D) = |x_2 - x_1|$$

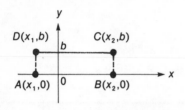

Figure 2

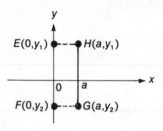

Figure 3

Similarly, in Figure 3,

$$d(E,F) = d(G,H) = |y_2 - y_1|$$

Let $P(1,1)$ and $Q(4,5)$ be two points in the plane (see Figure 4). Suppose we wish to find the length of the line segment PQ. We locate a point R so that PQR is a right triangle with PQ as the hypotenuse. One point is $R(4,1)$. Why? The distance from P to R, denoted by $d(P,R)$, is

$$d(P,R) = |4-1| = 3$$

and the distance from Q to R, denoted by $d(Q,R)$, is

$$d(Q,R) = |5-1| = 4$$

Therefore, using the Pythagorean theorem, we have

$$d(P,Q) = \sqrt{[d(P,R)]^2 + [d(Q,R)]^2}$$

$$= \sqrt{3^2 + 4^2}$$

$$= \sqrt{25} = 5$$

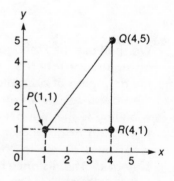

Figure 4

Now we will consider the general case.

THEOREM 1.1   Let $P(x_1,y_1)$ and $Q(x_2,y_2)$ be two points in R^2. Then

$$d(P,Q) = \sqrt{(x_2-x_1)^2+(y_2-y_1)^2}$$

Proof. Suppose P and Q do not lie on a horizontal or vertical line (see Figure 5). Select the point $T(x_2,y_1)$. Then the triangle with vertices P,Q, and T is a right triangle with PQ as the hypotenuse.

$$d(P,T) = |x_2-x_1|$$

$$d(Q,T) = |y_2-y_1|$$

Using the Pythagorean theorem, we have

$$[d(P,Q)]^2 = [d(P,T)]^2+[d(Q,T)]^2$$

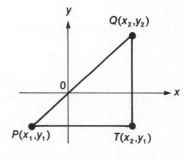

or
$$d(P,Q) = \sqrt{|x_2-x_1|^2+|y_2-y_1|^2}$$

$$= \sqrt{(x_2-x_1)^2+(y_2-y_1)^2}$$

Figure 5

as asserted.

If P and Q lie on a horizontal line, then $y_1=y_2$ and

$$d(P,Q) = |x_2-x_1| = \sqrt{(x_2-x_1)^2+(y_2-y_1)^2}$$

Similarly, if P and Q lie on a vertical line, then $x_1=x_2$ and

$$d(P,Q) = |y_2-y_1| = \sqrt{(x_2-x_1)^2+(y_2-y_1)^2}$$

This completes the proof.

EXAMPLE 1. Let $P(-1,-2)$ and $Q(3,4)$ be two points in the plane. Compute $d(P,Q)$.

Solution. We use Theorem 1.1 (see Figure 6).

$$d(P,Q) = \sqrt{[3-(-1)]^2+[4-(-2)]^2}$$

$$= \sqrt{4^2+6^2}$$

$$= \sqrt{52} = 2\sqrt{13}$$

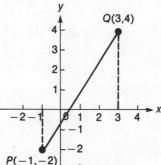

Figure 6 $P(-1,-2)$

It should be noted that it makes no difference whether the coordinates of P or those of Q are chosen as (x_1,y_1). It is easy to show that $d(P,Q)=d(Q,P)$.

EXAMPLE 2. A pilot is flying from Stannerville to Middletown to Bentow. With reference to a coordinate system, Stannerville, Middletown, and Bentow are located at $S(1,1)$, $M(4,5)$, and $B(6,13)$ respectively. (Assume that each unit equals 100 kilometers.)

    (a) Compute the distance traveled by the pilot.

    (b) If the pilot were to fly directly from Stannerville to Bentow, what is the distance traveled?

Solution. We use Theorem 1.1 (see Figure 7).

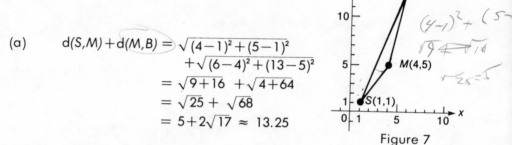

(a)
$$d(S,M)+d(M,B) = \sqrt{(4-1)^2+(5-1)^2}$$
$$+\sqrt{(6-4)^2+(13-5)^2}$$
$$= \sqrt{9+16} +\sqrt{4+64}$$
$$= \sqrt{25} + \sqrt{68}$$
$$= 5+2\sqrt{17} \approx 13.25$$

Figure 7

Therefore, the distance traveled by the pilot is approximately 1,325 kilometers.

(b) $d(S,B)= \sqrt{(6-1)^2+(13-1)^2}$
$$= \sqrt{25+144}$$
$$= \sqrt{169} = 13$$

Hence, the distance between Stannerville and Bentow is 1,300 kilometers.

EXAMPLE 3. Show that the triangle with vertices $A(-1,-1)$, $B(-2,6)$, and $C(1,0)$ is a right triangle.

Solution. We use Theorem 1.1.

$$d(A,B)=\sqrt{[-2-(-1)]^2+[6-(-1)]^2}$$
$$=\sqrt{1+49}$$
$$=\sqrt{50}$$
$$d(A,C)=\sqrt{[1-(-1)]^2+[0-(-1)]^2}$$
$$=\sqrt{4+1}$$
$$=\sqrt{5}$$
$$d(B,C)=\sqrt{[1-(-2)]^2+(0-6)^2}$$
$$=\sqrt{9+36}$$
$$=\sqrt{45}$$

Clearly,
$$(\sqrt{50})^2= (\sqrt{5})^2 + (\sqrt{45})^2$$

or
$$[d(A,B)]^2=[d(A,C)]^2+ [d(B,C)]^2$$

Therefore, by the converse of the Pythagorean theorem, which is, if a, b, and c are lengths of the sides of a triangle such that $a^2 + b^2 = c^2$, then the triangle is a right triangle, the triangle with vertices at A, B, and C is a right triangle with AB as the hypotenuse (see Figure 8).

Let $P(x_1, y_1)$ and $Q(x_2, y_2)$ be two points in the plane. We shall now show how we find the midpoint of the line segment PQ. Let $M(x, y)$ be the midpoint of the line segment PQ. We want to express x and y in terms of x_1, y_1, x_2, and y_2 (see Figure 9).

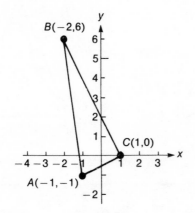

Figure 8

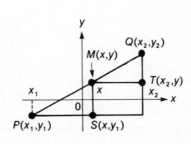

Figure 9

On the vertical line containing Q, consider the point $T(x_2, y)$, and on the horizontal line containing P, consider the point $S(x, y_1)$. The two triangles PMS and MQT are congruent. It follows that the corresponding sides and angles are equal. Hence,

$$d(P,S) = d(M,T)$$

Thus:

$$
\begin{aligned}
|x - x_1| &= |x_2 - x| \\
x - x_1 &= x_2 - x \qquad \text{(since } x > x_1 \text{ and } x_2 > x\text{)} \\
2x &= x_1 + x_2 \\
x &= \frac{x_1 + x_2}{2}
\end{aligned}
$$

Similarly, we find

$$y = \frac{y_1 + y_2}{2}$$

Therefore, the midpoint M on the line segment PQ is

$$M = \left(\frac{x_1 + x_2}{2}, \frac{y_1 + y_2}{2} \right)$$

EXAMPLE 4. Let $P(-2,1)$ and $Q(5,-3)$ be two points in the plane. Find the midpoint of the line segment PQ.

Solution. Using the midpoint formula we have

$$x = \frac{-2+5}{2} = \frac{3}{2}$$

$$y = \frac{1-3}{2} = -1$$

Therefore, the midpoint is $M(\frac{3}{2}, -1)$ (see Figure 10).

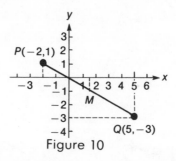

Figure 10

Self-test. Consider the two points $P(2,-3)$ and $Q(-1,5)$.

(a) $d(P,Q) = \underline{\hspace{2cm}}$

(b)   The midpoint of the line segment PQ has coordinates
$x = \underline{\hspace{2cm}}$ and $y = \underline{\hspace{2cm}}$.

NOTE: A triangle is said to be **isosceles** if the lengths of its two sides are equal. When all three sides are equal, it is known as an **equilateral** triangle.

The **median** of a triangle is the line segment drawn from the vertex of a triangle to the midpoint of the opposite side.

When three points lie in the same straight line, they are said to be **collinear**.

ANSWERS:

(a) $d(P,Q) = \sqrt{(2-(-1))^2 + (-3-5)^2} = \sqrt{9+64} = \sqrt{73}.$

(b) $x = (2+(-3))/2 = -\frac{1}{2}$, $y = (-3+5)/2 = 1.$

EXERCISES 1.2

In Exercises 1-12, plot each pair of points in a Cartesian coordinate system and find the distance between them. Then find the midpoint of the line segment joining the two points.

1. $(3,2),(1,-\frac{1}{2})$ 2. $(8,4),(3,1)$ 3. $(3,7),(0,5)$

4. $(4,-1),(5,2)$ 5. $(0,-2),(1,-3)$ 6. $(\frac{1}{2},2),(-3,\frac{1}{2})$

7. $(8,0),(-3,-4)$ 8. $(2,0),(6,3)$ 9. $(\frac{1}{2},\frac{1}{2}),(\frac{3}{2},-\frac{1}{2})$ *how do you determine order*

10. $(2,\frac{5}{2}),(-1,-\frac{3}{2})$ 11. $(0,\frac{7}{2}),(-\frac{7}{2},1)$ 12. $(5,-\frac{1}{2}),(-\frac{5}{2},-\frac{3}{2})$

In Exercises 13-18, plot each pair of points in a Cartesian coordinate system and show that triangle ABC is a right triangle.

13. $A(5,5)$, $B(-6,0)$, $C(2,-3)$

14. $A(2,3)$, $B(0,0)$, $C(6,-4)$

15. $A(4,1)$, $B(3,4)$, $C(-5,-2)$

16. $A(-2,3)$, $B(3,4)$, $C(5,-6)$

17. $A(2,-5)$, $B(-1,-3)$, $C(6,1)$

18. $A(3,-5)$, $B(-1,5)$, $C(-6,3)$

In Exercises 19-22, identify each triangle as **isosceles** or **equilateral**.

19. $A(5,-2)$, $B(6,-6)$, $C(2,-5)$

20. $A(-1,1)$, $B(1,-1)$, $C(-\sqrt{3},\sqrt{3})$

21. $A(5,5)$, $B(1,2)$, $C(4,-2)$

22. $A(5,0)$, $B(6,-4)$, $C(2,-3)$

23. If $M(-10,5)$ is the midpoint of AB and the coordinates of A are $(8,-3)$ find the coordinates of B.

24. If P is a point whose distance from $Q(3,0)$ is $\sqrt{65}$ and the abscissa of P is -5, find the ordinate of P. Is P unique?

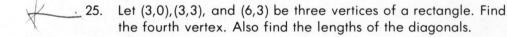

25. Let $(3,0),(3,3)$, and $(6,3)$ be three vertices of a rectangle. Find the fourth vertex. Also find the lengths of the diagonals.

26. Plot the points $A(4,3)$, $B(2,1)$, and $C(-2,-3)$ in a Cartesian coordinate system. Find $d(A,B)$, $d(A,C)$, and $d(B,C)$. Are the three points collinear? Explain

27. Let $A(5,-4)$, $B(7,3)$, and $C(2,1)$ be the vertices of a triangle. Find the length of each of the medians.

1.3 THE FUNCTION CONCEPT

OBJECTIVES

1. *Define a function.*
2. *Find the domain of a function.*
3. *Use a formula to find function values.*
4. *Graph the absolute value and greatest-integer functions.*

In the seventeenth century the methods of science were drastically altered. Prior to the seventeenth century, the scientific method was aimed at explanation rather than description of physical phenomena. Galileo*, among others, proposed that the methods of science be patterned after the methods of mathematics. In his studies on the motion of objects, Galileo came up with a formula of the form

$$d = 5t^2 \tag{1}$$

which describes how an object falls under the influence of gravity. This formula expresses algebraically the relationship between d, the distance an object falls

* GALILEO GALILEI

Galileo Galilei (1564-1642) was born in Pisa, Italy. He entered the University of Pisa to study medicine but switched to mathematics prior to graduating. He became a professor at Pisa and later at the University of Padua. In 1610, Grand Duke Casimo II de Medici appointed Galileo as chief mathematician at the court of Florence.

Galileo's interests went beyond mathematics. A talented writer, he published works in astronomy and physics as well as mathematics. In addition, he is credited with the discovery of the moons of the planet Jupiter, inventing a microscope, designing the first pendulum clock, and constructing and selling a popular compass.

*In 1633, the Roman Inquisition put Galileo under house arrest for his teaching of the Copernican system of astronomy. Although forbidden to publish, he continued to write, smuggling his book **Dialogues Concerning Two New Sciences** to Holland, where it was published in 1638.*

through space, and t, the time. The following table gives the relationship between certain values of t (measured in seconds) and the corresponding values of d (measured in meters).

t	0	1	2	3	4
d	0	5	20	45	80

Table 1

Studies of objects in motion led to the function concept. Among the outstanding men of the eighteenth century who contributed to the development of this concept are Leibniz, Huygens, and Euler. The word **function** was introduced into the mathematical language by Leibniz.

We all encounter functions in our everyday life:

> the association between the cost of a steak and
> the number of pounds it weighs

> the association between interest earned and
> the amount deposited into a savings account

> the association between the distance traveled and
> the speed at which one drives his car

are all examples of functions.

EXAMPLE 1. Let X be the set of students $\{s_1, s_2, \ldots, s_n\}$ in this algebra class and let Y be the set of grades $\{A, B, C, D, F\}$. Consider the relation in which each student in X is associated with a grade for the course. The important feature is that each student has only one grade for the course. We illustrate this in Figure 1. Here we say that if f is a rule that assigns to each student in X one and only one grade in Y, then f is called a **function**. The set X is called the **domain** of f. For each $s_i \in X$, the corresponding unique element in Y is called the **image** of s_i; the set of all images of the elements of X is called the **range** of the function f.

Note that a unique element in Y is associated with each $s_i \in X$. However, more than one element in X may be associated with the same element in Y.

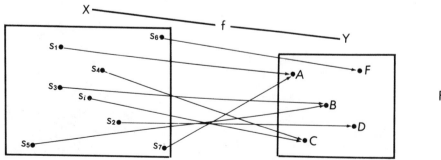

Figure 1

EXAMPLE 2. Two commonly used temperature scales are the **Celsius scale** and the **Fahrenheit scale**. We find that 0°C (zero degrees Celsius) is associated with 32°F (32 degrees Fahrenheit) and that 100°C is associated with 212°F. In general, the association between the elements on the Celsius scale and the elements on the Fahrenheit scale is given by the rule

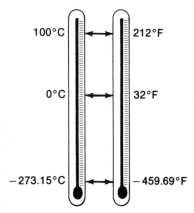

$$y = \tfrac{9}{5}x + 32 \qquad (2)$$

where the Celsius scale is represented by the set $X = \{x \mid x \geq -273.15\}$ and the Fahrenheit scale is represented by the set $Y = \{y \mid y \geq -459.67\}$. Thus, for each $x \in X$, there exists a unique $y \in Y$. Here we have a function f defined by Equation 2. The set X is the domain of f and the set Y is the range of f.

Figure 2

In the examples above one should note that we are dealing with
   (a)   two nonempty sets X and Y and
   (b)   a rule or formula that assigns to each element in X a single element in Y.

In Example 1, the association between the elements of the sets may be described by a finite set of ordered pairs, but it is not possible to list all of the ordered pairs in Example 2. However, Equation 2, together with a description of the sets X and Y defines a function.

We now give a formal definition of a function.

> **DEFINITION 1.2** A *function* is a nonempty set of ordered pairs of numbers (x,y) in which no two ordered pairs have the same first component. The set of all first components is called the *domain* of the function and the set of all second components is called the *range* of the function.

Note that all functions are relations but that not all relations are functions.

EXAMPLE 3. Which of the following relations is a function? Give the domain and range of each.

   (a) $f = \{(3,1),(1,2),(-1,3),(-3,5)\}$
   (b) $g = \{(2,4),(3,8),(2,-4),(0,-8)\}$
   (c) $h = \{(1,2),(2,3),(3,3),(4,5)\}$

Solution.

(a) This set of ordered pairs of numbers is a relation in which no two ordered pairs have the same first component. Therefore, f is a function.

  domain $f = \{-3,-1,1,3\}$
  range $f = \{1,2,3,5\}$

(b) The ordered pairs $(2,4)$ and $(2,-4)$ in g have the same first component. Therefore, the relation g is not a function.

  domain $g = \{0,2,3\}$
  range $g = \{-8,-4,4,8\}$

(c) This set of ordered pairs is a relation in which no two ordered pairs have the same first component. Therefore, the relation h is a function. Note that the ordered pairs $(2,3)$ and $(3,3)$ in h have the same second component; this does not alter the fact that h is a function.

  domain $h = \{1,2,3,4\}$
  range $h = \{2,3,5\}$

The following notation, which is due to Euler*, is commonly used. For each x in the domain of a function f, the corresponding value in the range is designated by $f(x)$, and we may write $y=f(x)$, read "y equals f of x". This denotes that y is a function of x. We call x the **independent variable** because it can take on any value in the domain, and we call y the **dependent variable** because its value depends upon our choice of x. The **graph** of a function f is the set of all points in R^2 such that (x,y) is an ordered pair in f.

In general, functions will be denoted by letters such as $f,g,h,F,$ and H. In Example 2, Equation 2 may be written as

$$f(x) = \tfrac{9}{5}x + 32$$

* LEONHARD EULER

Leonhard Euler (1707-1783) was a Swiss mathematician, receiving much of his early training from his father, who had studied mathematics with Jacob Bernoulli. Euler received his university education at Basel, majoring in theology. He received his Master's degree at age seventeen and wrote his first research paper two years later. Euler became one of the most prolific mathematicians of all times, writing more than 700 research papers in science.

During his lifetime Euler was appointed to the Chair of Mathematics at the Academy of Sciences in St. Petersburg, Russia, and served as the Director of the Department of Mathematics in Berlin. In 1735, he lost the sight in one eye and became totally blind some seventeen years before his death. The handicap notwithstanding, Euler continued to work to the end, calculating the orbit for the planet Uranus on the day he died.

Here, $f(x)$ plays the same role as y did in Equation 2. Thus, we write

$$f(0) = \tfrac{9}{5}(0) + 32 = 32$$
$$f(-10) = \tfrac{9}{5}(-10) + 32 = 14$$
$$f(-40) = \tfrac{9}{5}(-40) + 32 = -40$$

and so on, representing the value of y for a given value of x.

Self-test. Which of the following relations is a function? Give the domain and range of each.

(a) $f = \{(1,2),(2,3),(-1,3),(0,1)\}$ _____
    domain of $f =$ _____;
    range of $f =$ _____
(b) $g = \{(2,1),(0,2),(2,5),(-1,-5)\}$ _____
    domain of $g =$ _____;
    range of $g =$ _____

EXAMPLE 4. Graph the function f defined by
$$f(x) = 1 - 2x$$
if the domain of f is $\{-1,0,1,2,3\}$.

Solution. For the given values of x in the domain of f, we compute the corresponding values of $f(x)$. These values are listed in Table 2, and the graph of f is shown in Figure 3.

x	y = f(x)	(x, y)
−1	3	(−1, 3)
0	1	(0, 1)
1	−1	(1, −1)
2	−3	(2, −3)
3	−5	(3, −5)

Table 2

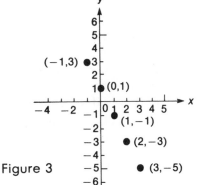

Figure 3

Self-test. If $f(x) = \frac{x}{3} - 1$,then

(a) $f(0)$ = _____

(b) $f(3)$ = _____

(c) $f(-1)$ = _____

(d) $f(\frac{1}{3})$ = _____

(e) $f(a)$ = _____

If the domain is not specified for a real function that is defined by a formula, then we shall assume that the domain consists of all real numbers for which the formula is meaningful. We illustrate this in the following example.

EXAMPLE 5. Describe the domain of each function defined by the following formulas.

(a) $f(x) = x^2$

(b) $g(x) = \frac{1}{x(x-1)}$

(c) $h(x) = \sqrt{x+2}$

(d) $f(x) = \frac{1}{\sqrt{x^2+x-2}}$

Solution.

(a) Since x^2 is a real number for every real number x, then domain $f = \{x \mid x \in R\}$.

(b) For this equation to be meaningful, we must have $x(x-1) \neq 0$. That is, $x \neq 0$ and $x \neq 1$. Therefore, domain $g = \{x \mid x \in R, x \neq 0, x \neq 1\}$.

(c) Since the square root of a number is defined when the number is greater than or equal to zero, the expression $\sqrt{x+2}$ is a real number if and only if $x+2 \geq 0$, that is, if $x \geq -2$. Thus, domain $h = \{x \mid x \geq -2\}$.

(d) Here $f(x)$ is a real number for values of x for which $x^2+x-2>0$. Therefore domain $f = \{x \mid x < -2\} \cup \{x \mid x > 1\}$.

ANSWERS:

(a) -1 (b) 0 (c) $-\frac{4}{3}$ (d) $-\frac{8}{9}$ (e) $\frac{a}{3}-1$

Self-test. Describe the domain of each function.

(a) $f(x) = \dfrac{1}{x-1}$ domain $f=$_____

(b) $g(x) = \dfrac{1}{x^2-4}$ domain $g=$_____

(c) $h(x) = \sqrt{x^2-1}$ domain $h=$_____

EXAMPLE 6. The **absolute value function** is defined by
$$f(x) = |x|$$
Describe the domain and range of f and sketch the graph of f.

Solution. Since for every real number $x, |x|$ is a real number, we have
$$\text{domain } f = \{x \,|\, x \in R\}$$
Since $|x| \geq 0,$
$$\text{range } f = \{y \,|\, y \geq 0\}$$

Using the formula $f(x) = |x|$, we construct Table 3. We note that for every $x > 0$, the point (x,x) belongs to the graph of f and for every $x < 0$, the point $(x, -x)$ belongs to the graph of f. From this we obtain the graph of f as shown in Figure 4.

x	y = f(x)	(x, y)
−2	2	(−2, 2)
−1	1	(−1, 1)
0	0	(0, 0)
1	1	(1, 1)
2	2	(2, 2)

Table 3

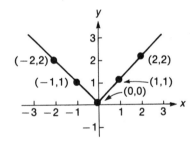

Figure 4

A function that is useful in computer programming and is of interest to the United States Postal Service is given in the following example.

EXAMPLE 7. The **greatest integer function** is defined by
$$f(x) = [\![x]\!]$$
where $[\![x]\!]$ represents the greatest integer that is less than or equal to x. Sketch the graph of f.

Solution: We list some of the values of f in Table 4. The graph is shown in Figure 5. Note that each line segment is open on the right end and closed on the left end. We note that the domain of f is R and the range of f is Z, the set of integers.

---

x	$y = f(x) = [x]$
$-3 \le x < -2$	-3
$-2 \le x < -1$	-2
$-1 \le x < 0$	-1
$0 \le x < 1$	0
$1 \le x < 2$	1
$2 \le x < 3$	2
$3 \le x < 4$	3

Table 4

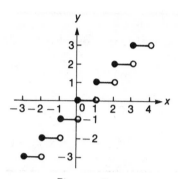

Figure 5

EXAMPLE 8. Is the relation defined by the equation $x^2+y^2=1$ a function?

Solution. Solving explicitly for y, we have
$$y=\sqrt{1-x^2} \quad \text{or} \quad y=-\sqrt{1-x^2}$$
We find that y is defined if $-1 \le x \le 1$. Now if $-1 < x < 1$, there are two values of y associated with each x. Therefore, the relation is not a function.

We note that a function involves two sets called the **domain** and the **range** and may be thought of as a function-machine with the following rules.

(a) The function-machine processes only elements in the domain and produces only elements in the range.

(b) A given input always produces the same output. The function-machine is not ambiguous.

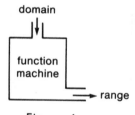

Figure 6

EXERCISES 1.3

In Exercises 1-6, determine which of the given sets of ordered pairs is a function. Give the domain and range of each.

1. $\{(1,3),(2,7),(3,11),(4,15)\}$
2. $\{(0,-1),(2,2),(3,-1),(4,5)\}$
3. $\{(4,2),(6,4),(8,6)\}$
4. $\{(1,-1),(2,7),(-1,1),(3,-1)\}$
5. $\{(0,8),(5,2),(3,8),(4,8)\}$
6. $\{(5,7),(2,-3),(3,7),(5,16)\}$

In Exercises 7-20, describe the domain of the function defined by each equation.

7. $f(x) = 2x - 1$

8. $f(x) = 2/(x + 1)$

9. $h(x) = x^2 + 1$

10. $g(x) = 1/(x^2 + 1)$

11. $F(t) = 4/(t^2 - t)$

12. $H(u) = \sqrt{u + 4}$

13. $h(x) = 1(x^2 - 1)$

14. $f(x) = \sqrt[3]{x + 1}$

15. $g(x) = \sqrt{4 - x^2}$

16. $g(t) = \sqrt{t^2 - t}$

17. $f(x) = (x^2 - 1)/(2x + 1)$

18. $f(x) = \sqrt{x^2 - 5x + 4}$

19. $g(x) = \sqrt[3]{1/(1 - x)}$

20. $h(x) = (\sqrt{x + 1})/(\sqrt{1 - x^2})$

21. If the function f is defined by $f(x) = x^2 - 1$, find each of the following.
(a) $f(0)$ (b) $f(1)$ (c) $f(-1)$ (d) $f(\frac{1}{2})$

22. If $f(x) = \sqrt{x - 1}$, find each of the following. What restrictions would you impose on the variable r in Part d?
(a) $f(1)$ (b) $f(5)$ (c) $f(\frac{5}{4})$ (d) $f(r + 1)$

In Exercises 23-25 graph each function.

23. $f(x) = 3x + 1$; domain $f = \{0,1,2,3,4\}$
24. $f(x) = 2 - x^2$; domain $f = \{-2, -1, 0, 1, 2, 3\}$
25. $h(x) = \sqrt{x + 1}$; domain $h = \{-1, 0, 1, 2, 3\}$

26. If the function f is defined by $f(x) = 3x + 4$, find a number in the domain of f such that:
(a) $f(x) = 0$ (b) $f(x) = 2$ (c) $f(x) = 4$

27. If the function g is defined by $g(x) = x^2 - 1$, find all numbers in the domain of g such that:
(a) $g(x) = 0$ (b) $g(x) = 3$ (c) $g(x) = 4$
Is there a number in the domain of g such that $g(x) = -2$? Explain.

28. If $f(x) = x^2$, is $f(a + b) = f(a) + f(b)$? For what values of a and b will the equality hold?

29. A function f is said to be **even** if $f(x)=f(-x)$ for all x and $-x$ in the domain of f; and f is said to be **odd** if $f(-x)=-f(x)$ for all x and $-x$ in the domain of f.
Which of the following functions are odd and which are even?
(a) $f(x)=x^2$ (b) $f(x)=x^3$ (c) $f(x)=x^4-x^2$ (d) $f(x)=x^3-x^2$

30. Give an example of a function which is both odd and even.

31. Give an example of a function which is neither odd nor even.

In Exercises 32-36, sketch the graph of each function.

32. $f(x) = \begin{cases} 2 & \text{if } x \le 1 \\ -1 & \text{if } x > 1 \end{cases}$

33. $g(x) = \begin{cases} 1 & \text{if } x \le 2 \\ x & \text{if } x > 2 \end{cases}$

34. $h(x) = \begin{cases} x & \text{if } -1 \le x \le 1 \\ 1 & \text{if } 1 < x \le \frac{5}{2} \\ 2 & \text{if } x > \frac{5}{2} \end{cases}$

35. $f(x) = |x-1|$

36. $f(x) = [\![x-1]\!]$

37. Suppose a worker is paid at a rate of \$4 per hour and receives time and a half for hours over 40 per week.
(a) Show that the pay as a function of time is described by the function f where

$$f(t) = \begin{cases} 4t & \text{if } t \le 40 \\ 160+6(t-40) & \text{if } t > 40 \end{cases}$$

(b) Sketch the graph of f.
(c) Compute $f(5)$, $f(10)$, $f(40)$, $f(50)$, and $f(60)$.

38. The equation $f(t)=5t^2$ describes the distance traveled by a falling object as a function of time. If the object is dropped from a building 180 meters high, find each of the following
(a) The distance traveled by the object in 1 second, 2 seconds, and 4 seconds.
(b) The time it takes the object to hit the ground.

39. The size of a picture on a screen is a function of the distance of the projector from the screen. If p is the size of the picture (in square feet) and x is the distance (in feet) from the projector to the screen, then it is found that
$$p(x) = x^2$$
Find $p(1)$, $p(2)$, $p(3)$, $p(4)$, and $p(5)$. Graph the function p for the given values of x.

40. A book club offers six books at $5.50 each and additional books are sold at half price with a limit of twelve books per customer.
    (a) Express the cost of the books as a function of the number of books sold per customer.
    (b) What is the cost of buying ten books?

41. The Bargain Car Rental Company charges $8.00 plus 12 cents per kilometer.
    (a) Express the rental cost as a function of the number of kilometers driven.
    (b) What is the cost of renting a car for a 75-km trip?

42. The owner of a pizza shop is selling 200 pizzas per day at the price of $3.00 per pizza. He estimates that for each 20 cent increase in the price, he will sell ten fewer pizzas per day. His fixed overhead is $150.00 per day plus 80 cents a pizza for ingredients.
    (a) Express the daily profit as a function of the selling price.
    (b) Compute the profit if the pizzas are sold at
        (i) $3.20, (ii) $3.60, (iii) $4.00.

1.4 OPERATIONS ON FUNCTIONS

OBJECTIVES

1. Define sum, difference, product, and quotient of two functions.
2. Define composition of functions.

The operations of addition, subtraction, multiplication, and division can be applied to functions. Let f and g be two functions whose domains are X_1 and X_2, respectively. Then the sum $f+g$, the difference $f-g$, the product fg, and the quotient f/g are defined as follows:

$$(f+g)(x) = f(x)+g(x) \tag{1}$$
$$(f-g)(x) = f(x)-g(x) \tag{2}$$
$$(fg)(x) = f(x)g(x) \tag{3}$$
$$(f/g)(x) = \frac{f(x)}{g(x)}, \quad g(x)\neq 0 \tag{4}$$

The domain of $f+g$, $f-g$, and fg is $X_1 \cap X_2$. The domain of f/g is $\{x \mid x \in X_1 \cap X_2$ and $g(x)\neq 0\}$. If $X_1 \cap X_2 = \emptyset$, then $f+g$, $f-g$, fg, and f/g are not defined.

EXAMPLE 1. If f and g are functions defined by

$$f(x)=\frac{1}{x-1} \text{ and } g(x)=\sqrt{x+1}$$

describe each of the following functions.

(a) $f+g$ (b) $f-g$
(c) fg (d) f/g
(e) g/f (f) gg

Solution. $f(x) = \dfrac{1}{x-1}$ $G(x) = \sqrt{x+1}$

(a) $(f+g)(x) = [1/(x-1)] + \sqrt{x+1} = [1+(x-1)\sqrt{x+1}]/(x-1)$

(b) $(f-g)(x) = [1/(x-1)] - \sqrt{x+1} = [1-(x-1)\sqrt{x+1}]/(x-1)$

(c) $(fg)(x) = [1/(x-1)]\sqrt{x+1} = (\sqrt{x+1})/(x-1)$

(d) $(f/g)(x) = [1/(x-1)] \div \sqrt{x+1} = 1/(x-1) \times 1/(\sqrt{x+1})$

$= 1/(x-1)(\sqrt{x+1})$

(e) $(g/f)(x) = \sqrt{x+1} \div 1/(x-1) = (x-1)(\sqrt{x+1})$

(f) $(gg)(x) = g^2(x) = (\sqrt{x+1})^2 = x+1$

The domain of f is $\{x \mid x \in R, x \neq 1\}$ and the domain of g is $\{x \mid x \geq -1\}$. Thus in Parts a, b, and c, the domain is $\{x \mid x \geq -1 \text{ and } x \neq 1\}$. In Part d, the domain is $\{x \mid x > -1 \text{ and } x \neq 1\}$; in Part e, the domain is $\{x \mid x \geq -1 \text{ and } x \neq 1\}$. Care should be exercised in Part f, since the domain of g^2 is $\{x \mid x \geq -1\}$. (Why?)

Self-test. If the functions f and g are defined by
$$f(x) = \sqrt{9-x^2} \text{ and } g(x) = 1-x^2$$
describe the following functions.

(a) $(f+g)(x) = $ _____
    domain of $f+g$: _____
(b) $(f-g)(x) = $ _____
    domain $(f-g)$: _____
(c) $(fg)(x) = $ _____
    domain (fg): _____
(d) $(f/g)(x) = $ _____
    domain (f/g): _____
(e) $(f^2)(x) = $ _____
    domain f^2: _____

Another useful operation on functions is the **composition** of functions.

DEFINITION 1.3 Let *f* and *g* be two functions. The *composite function* of *f* and *g*, denoted by *f*∘*g*, is defined by
$$(f∘g)(x)=f(g(x))$$
The domain of *f*∘*g* is
$$\{x \mid x \in (\text{domain } g) \text{ and } g(x) \in (\text{domain } f)\}.$$

Note that the notation *(f∘g)(x)* indicates that we first apply g to x and then apply f to g(x) (see Figure 1). It is also possible to find *(g∘f)(x)=g(f(x))*; this notation means to first obtain the image of x under f and then apply g to f(x).

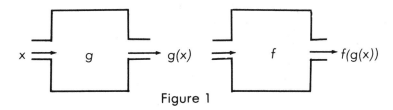

Figure 1

EXAMPLE 2. Let the functions *f* and *g* be defined by
$$f(x)=x^2+2 \quad \text{and} \quad g(x)=\sqrt{x-1}$$
Find *f*∘*g* and *g*∘*f*.

Solution. We use Definition 1.3.
$$\begin{aligned}
(f∘g)(x) &= f(g(x)) \\
&= f(\sqrt{x-1}) \\
&= (\sqrt{x-1})^2+2 \\
&= x-1+2 \\
&= x+1
\end{aligned}$$
We note that
$$\text{domain } f = \{x \mid x \in R\}$$
$$\text{domain } g = \{x \mid x \geq 1\}$$
Therefore,
$$\begin{aligned}
\text{domain } f∘g &= \{x \mid x \in (\text{domain } g) \text{ and } g(x) \in (\text{domain } f)\} \\
&= \{x \mid x \geq 1 \text{ and } \sqrt{x-1} \in R\} \\
&= \{x \mid x \geq 1\}
\end{aligned}$$

For *g*∘*f*, we have:
$$\begin{aligned}
(g∘f)(x) &= g(f(x)) \\
&= g(x^2+2) \\
&= \sqrt{(x^2+2)-1} \\
&= \sqrt{x^2+1}
\end{aligned}$$

domain $g \circ f = \{x \mid x \in (\text{domain } f) \text{ and } f(x) \in (\text{domain } g)\}$
$= \{x \mid x \in R \text{ and } x^2 + 2 \geq 1\}$
$= \{x \mid x \in R\}$

The next example illustrates how a complicated function may be expressed as the composition of simpler functions.

EXAMPLE 3. Let f be a function defined by $f(x) = \sqrt{x^2 + 5}$. Express f as the composition of two functions.

Solution. Let $g(x) = x^2 + 5$ and $h(x) = \sqrt{x}$. Then
$$(h \circ g)(x) = h(g(x))$$
$$= h(x^2 + 5)$$
$$= \sqrt{x^2 + 5}$$
$$= f(x)$$

Self-test. Let $f(x) = x^2$ and $g(x) = 2x - 1$. Find each of the following.

(a) $(f \circ g)(x) = $ _____

(b) $(g \circ f)(x) = $ _____

EXERCISES 1.4

In Exercises 1-6, functions f and g are defined. Describe $f+g$, $f-g$, fg, f/g, and g/f. Indicate the domain in each case.

1. $f(x)=3x+1$, $g(x)=2x^2$
2. $f(x)=2x^2+1$, $g(x)=\sqrt{x}$
3. $f(x)=x^2+2x-1$, $g(x)=\sqrt{x^2-1}$

4. $f(x)=\dfrac{1}{x+2}$, $g(x)=\dfrac{x}{x-1}$

5. $f(x)=\dfrac{1}{x^2}$, $g(x)=\dfrac{1}{\sqrt{x}}$

6. $f(x)=\sqrt{x^2-4}$, $g(x)=\sqrt[3]{x}$

In Exercises 7-14, describe $f\circ g$ and $g\circ f$ for the given functions of f and g. Indicate the domain in each case.

7. $f(x)=x+1$, $g(x)=2x$

8. $f(x)=2x^2-1$, $g(x)=4-x$

9. $f(x)=x^2+1$, $g(x)=\sqrt{x+1}$

10. $f(x)=2x^3+1$, $g(x)=\sqrt[3]{x}-1$

11. $f(x)=x-1$, $g(x)=1/(x+1)$

12. $f(x)=1/x$, $g(x)=x/(x^2+1)$

13. $f(x)=1/x^2$, $g(x)=\sqrt{x^2+3}$

14. $f(x)=x-1$, $g(x)=x^{-1/2}$

In Exercises 15-20, let $f(x)=\sqrt{x^2+4}$ and $g(x)=x^2$ and evaluate each expression.

15. $(f+g)(1)$ 16. $(f-g)(3)$ 17. $(fg)(-1)$
18. $(f/g)(\frac{1}{2})$ 19. $(f\circ g)(2)$ 20. $(g\circ f)(0)$

In Exercises 21-24, for each function f, describe $f \circ f$. Indicate the domain in each case.

21. $f(x) = 2x + 1$ 22. $f(x) = \dfrac{1}{x}$

23. $f(x) = \dfrac{1}{x} + 1$ 24. $f(x) = x + \dfrac{1}{x}$

25. If $f(x) = 2x + 1$, prove that $f(3x) - 3f(x) = -2$.

26. If $f(x) = x^2$, prove that $f(x+y) - f(x-y) = 4xy$ for every $x, y \in R$.

27. If $f(x) = 2x + 1$ and $g(x) = 3x + 2$, prove that $f \circ g = g \circ f$.

28. If $f(x) = x + 1$, $g(x) = 2x + 3$, and $h(x) = x^2$, find each of the following

   (a) $f(g(h(x)))$ (b) $g(h(f(x)))$ (c) $g(f(h(x)))$ (d) $h(g(f(x)))$

29. Let f be any function.
   (a) Show that $[f(x) + f(-x)]$ is an even function.
   (b) Show that $[f(x) - f(-x)]$ is an odd function.

30. Let f and g be odd functions.
   (a) Show that $f + g$ and $f - g$ are odd functions.
   (b) Show that fg and g/g are even functions.

1.5 INVERSE FUNCTIONS

Let *f* be a function. What happens if we interchange the components of every ordered pair in *f*? Is the resulting set of ordered pairs a function? For example, if

$$f=\{(1,0),(2,1),(3,1),(4,2)\}$$

then interchanging the components of these ordered pairs gives

$$\{(0,1),(1,2),(1,3),(2,4)\}$$

Clearly, the resulting set is not a function since the ordered pairs (1,2) and (1,3) have the same first component and different second components. On the other hand, for the function

$$f=\{(-1,2),(0,3),(5,-2),(-10,\tfrac{1}{2})\}$$

if we interchange the components of every ordered pair in *f*, we obtain a function *g*.

$$g=\{(2,1),(3,0),(-2,5),(\tfrac{1}{2},-10)\}$$

Now suppose a function *f* has no two ordered pairs with the same second component. Such a function is called a **one-to-one function** and interchanging the components of every ordered pair in *f* results in another function.

DEFINITION 1.4 A function *f* is said to be *one-to-one* if and only if, for every x_1 and x_2 in the domain of *f*,

$$f(x_1)=f(x_2) \text{ implies that } x_1=x_2$$

or equivalently,

$$x_1 \neq x_2 \text{ implies that } f(x_1) \neq f(x_2)$$

EXAMPLE 1. Is the function f defined by $f(x) = 3x - 1$ a one-to-one function?

Solution. Let x_1 and x_2 be real numbers. Then
$$f(x_1) = 3x_1 - 1 \quad \text{and} \quad f(x_2) = 3x_2 - 1.$$
Now, $f(x_1) = f(x_2)$ implies that
$$3x_1 - 1 = 3x_2 - 1$$
or
$$3x_1 = 3x_2$$
or
$$x_1 = x_2$$
Therefore, by Definition 1.4, f is a one-to-one function.

EXAMPLE 2. Is the function g defined by $g(x) = x^2$ a one-to-one function?

Solution. Here we find that
$$g(2) = 4 \quad \text{and} \quad g(-2) = 4$$
Thus, $g(2) = g(-2)$ even though $2 \neq -2$. Therefore, g is not a one-to-one function.

Self-test. Which of the following are one-to-one functions?

    (a) $f(x) = 2x^2 - 1$

    (b) $g(x) = 5x + 4$

If a function f is one-to-one, the relation g, obtained by interchanging the components of every ordered pair in f, is also a function. We call this function g the inverse function of f and we denote it by f^{-1} (read "f inverse").

DEFINITION 1.5 If f is a one-to-one function with domain X and range Y, then a function f^{-1} with domain Y and range X is called the *inverse function* of f if
$$f(f^{-1}(x)) = x \quad \text{for every } x \in Y$$
$$\text{and } f^{-1}(f(x)) = x \quad \text{for every } x \in X$$

EXAMPLE 3. Find f^{-1} given that
$$f = \{(-1,1),(-2,2),(-3,3),(-4,4)\}$$
Solution. It is clear that f is one-to-one. Therefore, interchanging the components of every ordered pair in f, we obtain
$$f^{-1} = \{(1,-1),(2,-2),(3,-3),(4,-4)\}$$
We note that
$$\text{domain } f^{-1} = \{1,2,3,4\} = \text{range } f$$
$$\text{range } f^{-1} = \{-1,-2,-3,-4\} = \text{domain } f$$

The graph of f and f⁻¹ are sketched on the same coordinate axes in Figure 1. An open circle is used to represent a point in f and an open square is used to represent a point in f⁻¹. Observe that the points (x,y) and (y,x) are symmetric with respect to the line $y=x$. Indeed, the graphs of f and f⁻¹ are symmetric with respect to the line $y=x$.

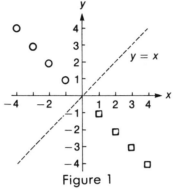

Figure 1

EXAMPLE 4. If f is the function defined by

$$f(x)=\tfrac{9}{5}\,x+32 \tag{1}$$

find the inverse function of f.

Solution. The student can readily show that f is a one-to-one function with domain R and range R. Hence the inverse function f⁻¹ exists. By Definition 1.5 we must have

$$f(f^{-1}(x))=x, \qquad \text{for every } x \in R$$

From Equation 1, we obtain

$$\tfrac{9}{5}(f^{-1}(x))+32=x, \qquad \text{for every } x \in R$$

Solving for f⁻¹(x), we get

$$f^{-1}(x)=\tfrac{5}{9}(x-32), \qquad \text{for every } x \in R \tag{2}$$

To verify that f⁻¹, as defined by Equation 2, is in fact the inverse function of f, we must check to see whether the conditions in Definition 1.5 are satisfied. Thus

$$f(f^{-1}(x))=f(\tfrac{5}{9}(x-32))$$
$$=\tfrac{9}{5}[\,\tfrac{5}{9}(x-32)]+32$$
$$=x$$

Also,

$$f^{-1}(f(x))=f^{-1}(\tfrac{9}{5}x+32)$$
$$=\tfrac{5}{9}[(\tfrac{9}{5}x+32)-32]$$
$$=x$$

Hence, Equation 2 defines the inverse function of f.

EXAMPLE 5. If f is a function defined by

$$f(x)=x^2-4, \qquad x \geq 0 \tag{3}$$

find the inverse function of f.

Solution. Since the domain of f is $\{x \mid x \geq 0\}$, it is easy to show that f is one-to-one. The range of f is $\{y \mid y \geq -4\}$. Let f^{-1} be the inverse function of f. Then, by Definition 1.5, we have

$$f(f^{-1}(x)) = x$$

or

$$[f^{-1}(x)]^2 - 4 = x$$

Solving for $f^{-1}(x)$, we obtain two possibilities:

$$f^{-1}(x) = \pm\sqrt{x+4}, \qquad x \geq -4 \tag{4}$$

If we let

$$f^{-1}(x) = \sqrt{x+4}$$

then

$$\begin{aligned} f(f^{-1}(x)) &= f(\sqrt{x+4}) \\ &= (\sqrt{x+4})^2 - 4 \\ &= (x+4) - 4 \\ &= x \end{aligned}$$

Also,

$$\begin{aligned} f^{-1}(f(x)) &= f^{-1}(x^2 - 4) \\ &= \sqrt{(x^2 - 4) + 4} \\ &= x \end{aligned}$$

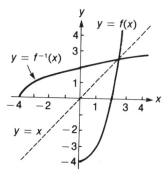

Figure 2

Therefore, Equation 4 defines the inverse function of f. If we select $f^{-1}(x) = -\sqrt{x+4}$, we find that $f(f^{-1}(x)) = x$. However, $f^{-1}(f(x)) = -x$ and so the expression $f^{-1}(x) = -\sqrt{x+4}$ does not define the inverse function of f. In Figure 2, we show the graphs of f and f^{-1} on the same coordinate axes. Notice the symmetric nature of the graphs.

Self-test.

   (a)   If $f = \{(2,10), (3,20), (4,30)\}$, then $f^{-1} = $ _____ .

   (b)   If g is a function defined by $g(x) = x^2$, $x \geq 0$, then $g^{-1}(x) = $ _____ .

ANSWERS:

(a) $f^{-1}(x) = \{(10,2),(20,3),(30,4)\}$.   (b) $g^{-1}(x) = \sqrt{x}$, $x \geq 0$.

EXERCISES 1.5

In Exercises 1-4, determine if f is a one-to-one function. If it is, find the inverse function.

1. $f=\{(0.5,3),(-5,7),(17,4)\}$
2. $f=\{(6,-2),(-2,6),(\frac{1}{2},\frac{1}{3})\}$
3. $f=\{(-3,1),(-2,0.5),(-\frac{1}{3},\frac{1}{2})\}$
4. $f=\{(10,1),(20,4),(-4,5),(-3,9)\}$
5. Show that the function f defined by $f(x)=x^2+1$ is not one-to-one.
6. Show that the function f defined by $f(x)=x^3$ is one-to-one.

In Exercises 7-16, find the inverse function f^{-1} for each function f.

7. $f(x)=7x+2$ 8. $f(x)=2x-3$
9. $f(x)=\sqrt{x+2}$ 10. $f(x)=\sqrt{x-2}$
11. $f(x)=3x^2, \ x\geq0$ 12. $f(x)=x^2+1, \ x\geq0$
13. $f(x)=x^3+1$ 14. $f(x)=\frac{1}{x}, \ x\neq0$
15. $f(x)=x^{1/3}+5$ 16. $f(x)=(x-2)^2$

17. Let $f(x)=15x-179$. Find each of the following.
    (a) $f^{-1}(f(5))$ (b) $f(f^{-1}(-135))$

18. Let $g(x)=x^3+4$. Find each of the following.
    (a) $g^{-1}(g(947))$ (b) $g(g^{-1}(-746))$

19. Use the same coordinate plane to sketch the graphs of f and f^{-1} defined in Exercise 8 above.

20. Use the same coordinate plane to sketch the graphs of f and f^{-1} defined in Exercise 10 above.

REVIEW OF CHAPTER ONE

1. Construct a number line.

2. Is there a one-to-one correspondence between the set of real numbers and the set of points on a line? Explain.

3. Is there a one-to-one correspondence between the set of integers and the set of points on a line? Explain.

4. Define the Cartesian product of two sets A and B.

5. If $P(-1,3)$ is a point in $R \times R$, then the abscissa of P is _____, the ordinate of P is _____ and P is in quadrant _____.

6. Let $A = \{x \mid |x+1| < 4\}$ and $B = \{-1,2\}$. Compute each of the following and give a geometric representation of each.
    $A \times B = $_____ $B \times A = $_____

7. In general, is $A \times B = B \times A$? Give an example in which $A \times B = B \times A$.

8. If $P(x_1,y_1)$ and $Q(x_2,y_2)$ are in R^2, then the distance between P and Q is $d(P,Q) = $_____.

9. If $P(-1,3)$ and $Q(2,-4)$ are in R^2, then $d(P,Q) = $_____.

10. What are the coordinates of the midpoint of the line segment joining $P(x_1,y_1)$ and $Q(x_2,y_2)$?

11. The coordinates of the midpoint of the line segment joining $P(4,-1)$ and $Q(1,3)$ are $x = $_____, $y = $_____.

12. Define a function.

13. What is the domain of function f where $f(x) = (\sqrt{x-2})/(3-x)$.

14. Explain the meaning of $[\![x]\!]$

In Exercises 15-20, let $f(x) = 1 - 2x^2$ and $g(x) = 1/(1-4x)$.

15. $f(-\frac{3}{2}) = $ _____

16. $g(a+h) = $_____

17. $(f+g)(x) = $_____

18. $(fg)(x) =$ _____

19. $(f/g)(0) =$ _____

20. $(g/f)(1) =$ _____

21. Define an even function.

22. Define an odd function.

23. Define the composite function of f and g.

24. Let $f(x) = \sqrt{1+x}$ and $g(x) = 1+x^2$.
    (a) $(f \circ g)(0) =$ _____
    (b) $(g \circ f)(0) =$ _____

25. Define a one-to-one function.

26. Is the function f defined by $f(x) = x^2$, $x < 0$, a one-to-one function? Explain.

27. Define inverse function.

28. If X and Y are the domain and range, respectively, of f, then domain $f^{-1} =$ _____ and range $f^{-1} =$ _____.

29. If $f(x) = 6x - 7$, then $f^{-1}(x) =$ _____.

30. If $g(x) = \sqrt{x+4}$, then $g^{-1}(x) =$ _____.

31. If f^{-1} is the inverse function of f, then $f(f^{-1}(u)) =$ _____ and $f^{-1}(f(v)) =$ _____.

TEST ONE

1. On a rectangular coordinate system locate the point $P(-3,5)$. The abscissa of A is _____ and the ordinate of A is _____. In which quadrant does A lie? _____

2. If $A=\{t \mid -2 \leq t \leq 3\}$ and $B=\{2\}$, then $A \times B =$_____. Graph $A \times B$.

3. Let $P(1,-2)$ and $Q(-3,4)$ be two points in the plane.
   (a) $d(P,Q)=$_____
   (b) The coordinates of the midpoint of the line segment PQ are
      $x=$_____ and $y=$_____.

4. Consider the set $h=\{(-5,1),(-4,2),(-3,3),(-2,4)\}$.
   (a) The domain of h is _____.
   (b) The range of h is _____.
   (c) Is h a function?_____. Explain.

5. If a function f is defined by $f(x)=3-\frac{x}{2}$, then
   (a) $f(0)=$_____
   (b) $f(-1)=$_____
   (c) $f(-2)=$_____

6. If a function g is defined by $g(x)=\dfrac{x}{x^2-5x+4}$, then the domain of g is _____.

7. If the function f is defined by $f(x)=\sqrt{2-x}$, then the domain of f is _____.

8. If $g(x)=1-x^2$, then for what values of x is $g(x)=-3$?

9. Let $f(x)=\sqrt{x-1}$ and $g(x)=\frac{1}{x}$. Then
   (a) $(f+g)(x)=$_____
   (b) $(f-g)(x)=$_____
   (c) $(fg)(x)=$_____
   (d) $(g/f)(3)=$_____
   (e) $(f/g)(2)=$_____

10. Let $f(x)=x^2+1$ and $g(x)=\sqrt{x}$. Then
    (a) $(f \circ g)(x)=$_____
    (b) $(g \circ f)(1)=$_____
    (c) $(g \circ g)(16)=$_____
    (d) $(f \circ f)(0)=$_____

11. Let $f(x)=1-2x$. Is f a one-to-one function?_____Explain.

12. Let $g(x)=\sqrt{x-4}$. Find $g^{-1}(x)$.

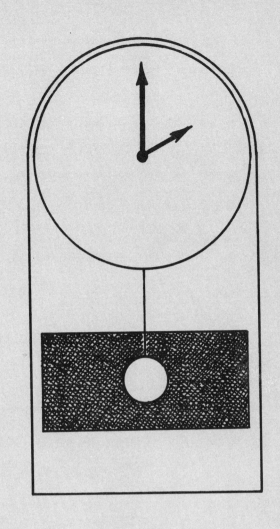

CHAPTER TWO:
THE TRIGONOMETRIC
FUNCTIONS

2.1 ANGLES AND THEIR MEASUREMENT

OBJECTIVES

1. *Define degree and radian measure of an angle.*
2. *Convert from degree to radian and from radian to degree measure.*
3. *Define coterminal angles.*
4. *Apply the formula $\theta = s/r$, which relates the length of an arc of a circle, its radius, and the measure of the central angles subtended by the arc.*

The branch of mathematics called trigonometry (from Greek words meaning "measuring of angles") is another example of the interaction between mathematics and social development. The ancient Alexandrian Greeks' creation of trigonometry was motivated by their desire to predict the paths and positions of the heavenly bodies and to aid in telling time, navigation, and geography. Trigonometry is now a branch of mathematics concerned with the properties and applications of the **circular** or **trigonometric functions**. Trigonometry continues to have extensive use in astronomy, surveying, and navigation, as well as engineering and the sciences in general.

A basic concept in trigonometry is the notion of an **angle**. When a half-line (or ray) is rotated in a plane about its end-point, an angle is generated (see Figure 1). The endpoint is called the **vertex** of the angle, the initial position of the half-line is called the **initial side**, and its final position is called the **terminal side** of the angle. An angle is said to be in **standard position** if the vertex is at the origin and the initial side is along the positive x-axis of a rectangular coordinate system (see Figure 2).

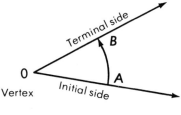

Figure 1

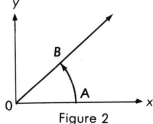

Figure 2

In the case of a line segment, we assign such measures as centimeters or inches, meters or yards; and in the case of areas, we assign measures such as square meters or square yards. To measure an angle, a basic unit of measure is needed. We assign a measure of an angle by means of a circle with its center at the vertex of the angle. There are two commonly used units of angle measure: the **degree** and the **radian**. The former was inherited from the ancient world of the Babylonians.

In geometry, the measure of an angle is always regarded as positive. In trigonometry however, the direction of rotation of the initial side is taken into consideration when assigning a measure to an angle. By convention, we assign positive numbers when the initial side is rotated in the **counterclockwise** direction and **negative** number when the initial side is rotated in the **clockwise** direction.

In order to define the degree measure, let us consider the set of all points one unit from the origin. Such a set of points is called the **unit circle**. The student can readily show by use of the distance formula that the equation of such a circle is $x^2+y^2=1$ (see Figure 3).

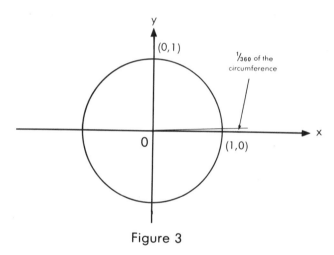

Figure 3

We assign the number 360 degrees (denoted by 360°) to the angle generated by one complete counterclockwise rotation of the initial side. Thus, we obtain an angle of **one degree measure** if the central angle in a circle is subtended by an arc 1/360 of the circumference of the circle. There are 60 minutes (written 60') in a degree and 60 seconds (written 60") in a minute. That is

$$360° = 1 \text{ complete rotation}$$
$$1° = 60'$$
$$1' = 60''$$

It is customary, when using calculators, to represent minutes and seconds as decimal fractions of degrees. Thus for example,

$$72°15' = (72+15/60)° = (72+0.25)° = 72.25°$$

In scientific work, a useful system for measuring angles is the **radian** measure. We recall from geometry that the circumference of a circle of radius r is $2\pi r$. Thus, for the unit circle, the circumference is 2π. We define radian measure as follows:

> **DEFINITION 2.1**   Let θ be an angle in standard position (see Figure 4). Let C be a circle of radius r and the center at the origin intersecting the initial and terminal sides of θ at A and B respectively. If s is the length of the arc AB, then the radian measure of θ is given by

$$\theta = s/r \tag{1}$$

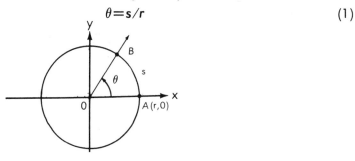

Figure 4

Note that s is considered negative if the angle is formed by a clockwise rotation. Also, it should be noted that θ represents both the angle and the measure of the angle. It is easy to see that the radian measure of an angle θ generated by a complete rotation (counterclockwise) is

$$\theta = 2\pi r/r = 2\pi \text{ radians}$$

Also note that when $s=r$, then the measure of the corresponding angle is 1 radian.

We now look at the relation between the degree measure and the radian measure of an angle. Since, by definition, one complete counterclockwise rotation of the terminal side generates an angle whose measure is 360° and also 2π radians, we have the relationship

$$360° = (2\pi)(\text{radians}) \tag{2}$$

Hence,

$$1° = \pi/180 \approx 0.01745 \text{ radians} \tag{3}$$

and

$$1 \text{ radian} = 180/\pi \approx 57.2958° \tag{4}$$

It is customary to omit the unit **radian**. Thus we write $90° = \pi/2$. In Table 1, we have the correspondence between degree and radian measures of some common angles.

Degrees	0°	15°	30°	45°	60°	90°	120°	135°	150°	180°	210°	225°	240°	270°	300°	315°	330°	360°
Radians	0	$\pi/12$	$\pi/6$	$\pi/4$	$\pi/3$	$\pi/2$	$2\pi/3$	$3\pi/4$	$5\pi/6$	π	$7\pi/6$	$5\pi/4$	$4\pi/3$	$3\pi/2$	$5\pi/3$	$7\pi/4$	$11\pi/6$	2π

Table 1

EXAMPLE 1. What is the radian measure of an angle measuring 56°45′21″?

Solution. First we note the following.

$$56°45′21″ = 56 + 45/60 + 21/3600 \text{ degrees}$$
$$\approx 56 + 0.75 + 0.0058 \text{ degrees}$$
$$\approx 56.7558°$$

Now, 1° = π/180. Thus we have

$$56.7558° = \cancel{180}^{\pi/180}\pi \times 56.7558° \approx 0.9906$$

EXAMPLE 2. What is the degree measure of an angle measuring 2.36 radians?

Solution. Since 1 radian ≈ 57.2958°, we have the following.

$$2.36 \text{ radians} \approx 57.2958 \times 2.36 \approx 135.218° \approx 135°13′4.8″$$

---

* **Self-test.** Fill in the blanks.
  (a) 73° = _____radians
  (b) 26°15′ = _____radians
  (c) 1.46 radians = _____degrees

EXAMPLE 3. Find the radian measure of a central angle subtended by an arc 15 cm long on a circle with a radius of 4 cm.

Solution. Substituting in Equation 1 we get

$$\theta = 15/4 = 3.75$$

Thus, the angle measure is 3.75 radians.

** **Self-test.** Compute the radian measure of a central angle
  (a) subtended by an arc 30 cm long on a circle with a radius of 8 cm. θ = _____
  (b) subtended by an arc 15 meters long on a circle with a radius of 4 meters. θ = _____

It should be noted that the radian measure of an angle is independent of the size of the circle.

ANSWERS:

*(a) 1.274   (b) 0.458   (c) 83.652   **(a) 3.75   (b) 3.75

2.4 GRAPHS OF THE SINE AND COSINE

OBJECTIVES

1. Graph the sine and cosine functions.
2. Graph $y=A\ sin(B\theta+C)$ and $y=A\ cos(B\theta+C)$ for different values of A,B, and C.
3. Determine the amplitude, cycle, and phase shift of sine waves.
4. Graph trigonometric functions by the addition of ordinates.

When we graph trigonometric functions, we may regard the independent variable as a real number. Thus, for the function given by $y=\sin\theta$, the number θ represents the degree or radian measure of an angle and is associated with the measure of a line segment on the x-axis of a rectangular coordinate system. Having agreed on this definition of the variable θ, we note that the domain of the sine function is the set of real numbers R and the range is the interval $[-1,1]$.

In this section, we shall construct the graphs of the sine and cosine functions. First we shall consider the sine function. Since the sine function is periodic, with a period of 2π, the graph of this function repeats the basic shape obtained on the interval $[0,2\pi]$. We can obtain a good approximation of the graph of the sine function by plotting a number of points $(\theta, \sin\theta)$ as θ takes values from 0 to 2π. We use Table 1 in Section 2.3 to evaluate $\sin\theta$. Next, we assume that the graph of the sine function is continuous (that is, it contains no breaks or gaps) and join these points with a smooth curve as shown in Figure 1. Then, because the sine function is periodic, the pattern shown in Figure 1 repeats itself over intervals with a length of 2π both to the left and to the right indefinitely. Thus, we obtain the graph of the sine function as shown in Figure 2.

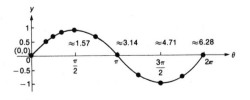

Figure 1

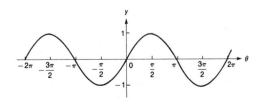

Figure 2

The student should learn where the maximum and minimum values of the function are, as well as the points where the graph crosses the θ-axis. Some of the points where the function attains a maximum value are

$$(-3\pi/2,1),\ (\pi/2,1),\ \text{and}\ (5\pi/2,1)$$

Some of the points where the function attains a minimum value are

$$(-5\pi/2,-1),\ (-\pi/2,-1),\ \text{and}\ (3\pi/2,-1)$$

The zeros of the sine function are $\theta=n\pi$, $n\in Z$(the set of integers).

The portions of the graph over any fundamental period of the function is called a **cycle** (or **wavelength**) of the curve. Half the difference between the maximum and minimum values of the function is called the **amplitude** of the curve. Thus, for $y=\sin\theta$, the amplitude is $\frac{1}{2}[1-(-1)]=1$.

The graph of $y=\cos\theta$ can be obtained in a similar manner as the graph of $y=\sin\theta$. Using Table 1 from Section 2.3, we plot a number of points $(\theta,\cos\theta)$. Connecting these points with a smooth curve, we obtain an approximate graph of the cosine function as shown in Figure 3.

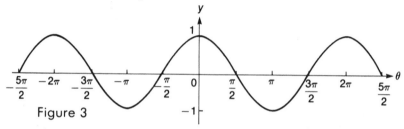

Figure 3

We observe that the graph of the cosine function is exactly the same as that of the sine function, except that the curve is moved $\pi/2$ units to the left on the coordinate system. The graphs of the sine and cosine functions are called **sine waves** (or **sinusoids**).

Self-test. Answer the following questions for the cosine function, $y=\cos\theta$.

(a) The graph crosses the y-axis at _____

(b) The amplitude is _____

(c) The zeros are at $\theta=$ _____

We shall now study functions defined by equations of the form
$$y=A\sin(B\theta+C)$$
and
$$y=A\cos(B\theta+C)$$
where A, B, and C are constants and A and B are not zero. The graphs of such functions are always sine waves. The values of B and C determine the cycle and $|A|$ is the amplitude of the curve.

ANSWERS:

(a) (0,1)   (b) 1   (c) $\theta=(2n+1)\pi/2$ whre n is any integer.

Sinusoidal waves occur in astronomy, mathematics, engineering and all of the sciences. The reader is undoubtedly familiar with the term **kilocycles per second** which is the radio frequency of the numbers on a radio dial, or the "60-cycle" electrical current used in a home. These represent the frequency of an oscillation, that is, the number of periods (or cycles) per unit of time.

In the study of musical tones, after a tuning fork is struck, its oscillation follows closely the function $y = A \sin B\theta$. The **loudness** of the sound produced by the fork depends on A and the **pitch** is determined by the frequency of oscillation.

Sinusoidal waves also occur in the study of heartbeats, as well as in the use of radar by police to measure the speed of cars.

Now we compare the graphs of $y = \sin \theta$ and $y = A \sin \theta$.

EXAMPLE 1. Sketch the graphs of $y = \sin \theta$ and $y = 2 \sin \theta$ on the same axes and to the same scale on $[0, 2\pi]$.

Solution. First we sketch the graph of $y = \sin \theta$ on $[0, 2\pi]$. For the function $y = 2 \sin \theta$, it is clear that for each $\theta \in [0, 2\pi]$, the ordinate is twice the corresponding ordinate of the graph of $y = \sin \theta$. We obtain the graph of $y = 2 \sin \theta$ on $[0, 2\pi]$ as shown in Figure 4. Note that the fundamental period of $y = 2 \sin \theta$ is the same as the fundamental period of $y = \sin \theta$.

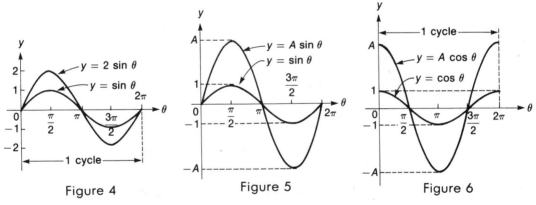

Figure 4   Figure 5   Figure 6

In Figure 5, we show the graph of $y = A \sin \theta (A > 1)$ as compared to the graph of $y = \sin \theta$. The graph of $y = A \cos \theta$ can be obtained in a similar fashion. In Figure 6, we compare the graphs of $y = \cos \theta$ and $y = A \cos \theta (A > 1)$ on $[0, 2\pi]$.

Self-test. State the amplitude of each graph.

    (a) $y = -3 \sin \theta$: _____

    (b) $y = \frac{1}{4} \cos \theta$: _____

    (c) $y = A \sin \theta$: _____

    (d) $y = A \cos \theta$: _____

ANSWERS: (d) $|A|$ (c) $|A|$ (b) $\frac{1}{4}$ (a) 3

Next we compare the graphs of $y=\sin \theta$ and $y=\sin B\theta$.

EXAMPLE 2. Sketch the graphs of $y=\sin \theta$ and $y=\sin 3\theta$ on the same axes and to the same scale on $[0,2\pi]$.

Solution. We know that the graph of $y=\sin \theta$ makes one complete cycle on $[0,2\pi]$. Now 3θ goes from 0 to 2π as θ goes from 0 to $\frac{2}{3}\pi$. Therefore, the graph of $y=\sin 3\theta$ makes one full cycle on $[0,\frac{2}{3}\pi]$; that is, its period is $\frac{2}{3}\pi$. Thus, on $[0,2\pi]$, the graph of $y=\sin 3\theta$ makes three cycles. The graphs of $y=\sin \theta$ and $y=\sin 3\theta$ on $[0,2\pi]$ are shown together in Figure 7.

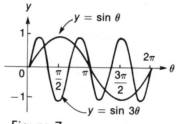

Figure 7

Similarly, the graph of $y=\sin 4\theta$ makes one complete cycles on $[0,\pi/2]$ and the graph of $y=\sin \frac{1}{2}\theta$ makes one complete cycle on $[0,4\pi]$. In general, for each positive number B, the graph of $y=\sin B\theta$ makes one complete cycle on $[0,2\pi/B]$. The amplitude of $y=\sin B\theta$ is 1 and its period is $2\pi/B$.

Note that the statements we have made about $y=\sin B\theta$ hold equally for $y=\cos B\theta$.

$\text{such that} \quad (\text{THIS}) \times (\text{Period}) = 2\pi$

Self-test. Give the amplitude and period for each of the following.

(a) $y=2 \sin 3\theta$: _____; _____

(b) $y=\frac{1}{3} \cos 4\theta$: _____; _____

(c) $y=3 \cos \frac{1}{4}\theta$: _____; _____

(d) $y=-2 \cos \frac{1}{2}\theta$: _____; _____

The graph of $y=-2 \cos \frac{1}{2}\theta$ on $[0,4\pi]$ is shown in Figure 8.

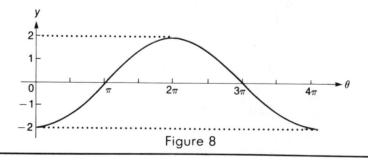

Figure 8

ANSWERS:

(a) $2, \frac{2}{3}\pi$ (b) $\frac{1}{3}; \pi/2$ (c) $3; 8\pi$ (d) $2; 4\pi$

Now we consider the graph of the functions of the form $y=\sin(\theta+C)$.

EXAMPLE 3. Sketch the graph of the function $y=\sin(\theta+\pi/4)$ on $[-2\pi,2\pi]$.

Solution. Using some of the special angles considered earlier, we construct the following table.

θ	0	$\pi/4$	$\pi/2$	$3\pi/4$	π	$5\pi/4$	$3\pi/2$	$7\pi/4$	2π
$\theta+\pi/4$	$\pi/4$	$\pi/2$	$3\pi/4$	π	$5\pi/4$	$3\pi/2$	$7\pi/4$	2π	$9\pi/4$
$\sin(\theta+\pi/4)$	0.707	1.000	0.707	0.000	-0.707	-1.000	-0.707	0.000	0.707

Table 1

We plot this data and connect the points with a smooth curve. We then extend the curve to $-2\pi\le\theta\le2\pi$ as shown in Figure 9.

Let us compare the graphs of $y=\sin(\theta+\pi/4)$ and $y=\sin\theta$. The graph of the former is a sine wave which is the graph of $y=\sin\theta$, shifted to the left by the amount of $\pi/4$.

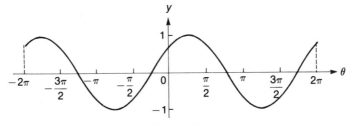

Figure 9

In general, the graph of $y=\sin(\theta+C)$ is a sine wave shifted to the left by an amount C if $C>0$ (see Figure 10). If $C<0$, the graph will be shifted to the right by an amount $|C|$. The constant C is called the **phase shift** or **phase angle**.

The most general sine wave is given by
$$y=A\,\sin(B\theta+C)$$
We note that when $B\theta+C=0$, $\theta=-C/B$, and when $B+C=2\pi$, $\theta=(2\pi-C)/B$. Thus, the graph of $y=A\sin(B\theta+C)$ is a sine wave with amplitude $|A|$, a period $2\pi/B$ and phase shift $-C/B$ (see Figure 11).

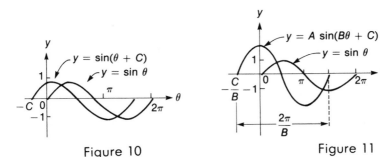

Figure 10   Figure 11

Self-test. Answer each question for the function $y=-2 \sin(\tfrac{1}{3}\theta + \pi/4)$

(a)   The graph is a _____

(b)   Amplitude: _____

(c)   Period: _____

(d)   Phase shift: _____

Note that everything we have said about $y=A \sin(B\theta+C)$ holds equally for
$$y=A \cos(B\theta+C).$$

Finally, we consider the graphing of functions expressed as the sum of two or more simpler functions. The graphing of such functions is best done by the method called the **addition of ordinates**. We illustrate this in the following example.

EXAMPLE 4. Graph the function $y=\cos 2\theta + \sin \theta$ on $[0, 4\pi]$.

Solution. First we draw the graphs of $y=\cos 2\theta$ and $y=\sin \theta$ on the same coordinate system over $[0, 4\pi]$. The ordinate of the graph $y=\cos 2\theta + \sin \theta$ at each point θ is the sum of the corresponding ordinates of $y=\cos 2\theta$ and $y=\sin \theta$ for the given θ. Thus, for $\theta=\pi/3$, $\sin \theta=0.866$, $\cos 2\theta=-0.500$, and so, $\sin \theta+\cos 2\theta=0.366$.

We do this for a few values of θ to obtain an approximation of the graph of $y=\cos 2\theta + \sin \theta$, as shown in Figure 12.

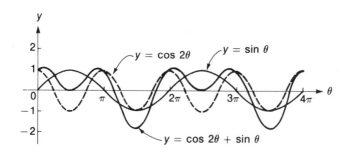

Figure 12

EXERCISES 2.4

In Exercises 1-30, sketch the graph of each function on $[-2\pi, 2\pi]$. Indicate the amplitude, period, and phase shift. When graphing these functions it will be helpful to use your knowledge of the general character, period, and amplitude of the curve.

1. $y = 2\cos\theta$ 2. $y = 3\sin\theta$ 3. $y = -2\cos\theta$

4. $y = \sin 4\theta$ 5. $y = \frac{1}{2}\sin 2\theta$ 6. $y = \frac{1}{2}\cos\theta$

7. $y = -3\sin\frac{1}{2}\theta$ 8. $y = -2\cos 4\theta$ 9. $y = \cos\frac{1}{2}\theta$

10. $y = -3\cos 2\theta$ 11. $y = \frac{1}{4}\sin\theta$ 12. $y = \sin(\theta/4)$

13. $y = 2\cos(\theta/4)$ 14. $y = \frac{1}{2}\cos 4\theta$ 15. $y = \sin(-3\theta)$

16. $y = -3\sin\theta$ 17. $y = -\frac{1}{2}\cos 3\theta$ 18. $y = -\frac{1}{2}\cos 2\theta$

19. $y = \sin(\theta + \pi)$ 20. $y = \cos(\theta + \pi)$ 21. $y = 2\cos(\theta - \pi/2)$

22. $y = 2\sin(\theta - \pi/2)$ 23. $y = \sin(\theta - \pi/4)$ 24. $y = 2\cos(\theta - \pi/4)$

25. $y = 3\sin(\theta + \pi/6)$ 26. $y = 3\cos(\theta + \pi/3)$ 27. $y = 2\sin(2\theta - \pi/6)$

28. $y = 3\cos(2\theta - \pi/6)$ 29. $y = 3\cos(\frac{1}{2}\theta - \pi/2)$ 30. $y = \frac{1}{2}\sin(2\theta + \pi/3)$

In Exercises 31-40, use the method of addition of ordinates to sketch a graph of each function on $[0, 4\pi]$.

31. $y = \sin\theta + \cos\theta$ 32. $y = 2\sin\theta - \cos\theta$ 33. $y = \sin 2\theta + \cos\theta$

34. $y = 2\sin\theta + \cos\theta$ 35. $y = 1 + \sin\theta$ 36. $y = 4 + 3\cos\theta$

37. $y = -3 + 2\sin\theta$ 38. $y = \sin\theta + \frac{1}{3}\sin 3\theta$ 39. $y = \theta - \sin\theta$

40. $y = \theta + \cos\theta$

2.5 OTHER TRIGONOMETRIC FUNCTIONS

OBJECTIVES

1. Define the tangent, cotangent, secant, and cosecant functions.
2. Prove some basic properties of these functions.
3. Find the function values of some special angles.
4. Find the five function values, given the sixth function value of an angle and the quadrant in which the terminal side is.
5. Sketch the graphs of these functions.

The sine and the cosine are the basic trigonometric functions. Certain combinations of these functions occur so often that they are given special names. There are four other trigonometric functions which are defined as follows.

DEFINITION 2.5   Let θ be any angle in the standard position and let (x,y) be any point (except the origin) on the terminal side of θ. Then the *tangent, cotangent, secant,* and *cosecant* functions are defined by

$\tan \theta = y/x$ (read "tangent of θ")
$\cot \theta = x/y$ (read "cotangent of θ")
$\sec \theta = r/x$ (read "secant of θ")
$\csc \theta = r/y$ (read "cosecant of θ")

where $r=\sqrt{x^2+y^2}$.

In each case, the denominator must be different from zero. Thus, the tangent and secant functions are not defined on the set $A=\{\theta\,|\,\theta=[(2k+1)/2]\pi, k\in Z\}$, since for any $\theta\in A$, the terminal side is on the y-axis and so $x=0$. Therefore, the domain of the tangent and secant functions is the set $R-A$ (the set of numbers that are in R but not in A).

Similarly, the cotangent and cosecant functions are not defined on the set $B=\{\theta\,|\,\theta=k\pi, k\in Z\}$. Therefore, the domain of the cotangent and cosecant functions is the set $R-B$.

From Definition 2.5, we note that these four trigonometric functions can be expressed in terms of the sine and cosine. Thus, for example,

$$\tan\theta = y/x = (y/r)/(x/r) = \sin\theta/\cos\theta, \ \cos\theta \neq 0 \tag{1}$$

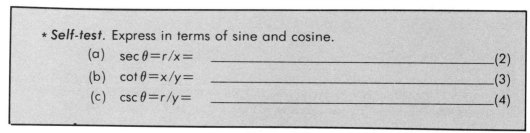

* **Self-test.** Express in terms of sine and cosine.

    (a) $\sec\theta = r/x =$ _____(2)

    (b) $\cot\theta = x/y =$ _____(3)

    (c) $\csc\theta = r/y =$ _____(4)

For the tangent function, we note that

$$\tan(-\theta) = \sin(-\theta)/\cos(-\theta) = -\sin\theta/\cos\theta = -\tan\theta \tag{5}$$

Thus, the tangent function is an odd function. Also,

$$\tan(\theta+\pi) = \sin(\theta+\pi)/\cos(\theta+\pi) = -\sin\theta/-\cos\theta = \tan\theta \tag{6}$$

Thus, the tangent function is a periodic function with period π.

** **Self-test.** Using similar arguments as above, answer the following.

    (a) Is the cotangent function even or odd?

    (b) Is the secant function even or odd?

    (c) Is the cosecant function even or odd?

    (d) The period of $\cot\theta$ is _____

           of $\sec\theta$ is   _____

           and of $\csc\theta$ is _____

As in the case of the sine and cosine functions, the signs of the other four trigonometric functions depend upon the quadrant in which the terminal side of θ lies, when in the standard position. Thus, we see that $\tan\theta = y/x$ is positive when the terminal side of θ is in the first and third quadrants and $\tan\theta$ is negative when the terminal side of θ is in the second and fourth quadrants. In Table 1, we give the signs of $\tan\theta$, $\cot\theta$, $\sec\theta$, and $\csc\theta$ for the terminal side of θ in the different quadrants.

Quadrant	I	II	III	IV
$\tan\theta$	+	−	+	−
$\cot\theta$	+	−	+	−
$\sec\theta$	+	−	−	+
$\csc\theta$	+	+	−	−

Table 1

ANSWERS:

** (a) Odd   (b) Even   (c) Odd   (d) The period of $\cot\theta$ is π, and of $\sec\theta$ and $\csc\theta$ is 2π.

* (a) $1/(x/r) = 1/\cos\theta$, $\cos\theta \neq 0$ (b) $(x/r)/(y/r) = \cos\theta/\sin\theta$, $\sin\theta \neq 0$ (c) $1/(y/r) = 1/\sin\theta$, $\sin\theta \neq 0$.

In the following example, we give two useful identities.

EXAMPLE 1. Prove each of the following identities.
$$1+\tan^2\theta=\sec^2\theta, \ \cos\theta\neq0 \tag{7}$$
$$1+\cot^2\theta=\csc^2\theta, \ \sin\theta\neq0 \tag{8}$$

Solution. We shall prove Identity 7. The proof of Identity 8, which is very similar to the proof of Identity 7, will be left as an exercise for the student.
We have

$$
\begin{aligned}
1+\tan^2\theta &= 1+(\sin^2\theta)/(\cos^2\theta), && \text{provided } \cos\theta\neq0\\
&= (\cos^2\theta+\sin^2\theta)/(\cos^2\theta), && \text{provided } \cos\theta\neq0\\
&= 1/(\cos^2\theta), && \text{provided } \cos\theta\neq0\\
&= \sec^2\theta, && \text{provided } \cos\theta\neq0
\end{aligned}
$$

This completes the proof.

The values of $\tan\theta$, $\cot\theta$, $\sec\theta$, and $\csc\theta$ for the special angles we considered in Section 2.3 can be easily determined.
For example, for $\theta=\pi/4$, we find
$$\tan\pi/4=(\sin\pi/4)/(\cos\pi/4)=(\sqrt{2}/2)/(\sqrt{2}/2)=1$$
and
$$\sec\pi/4=1/(\cos\pi/4)=1/(\sqrt{2}/2)=\sqrt{2}$$
and so on.

For $\theta=\pi/2$, we find
$$\tan\pi/2=(\sin\pi/2)/(\cos\pi/2)$$
which is not defined, since $\cos\pi/2=0$. Similarly, $\sec\pi/2$ is not defined. In Table 2, we give the values of $\tan\theta$, $\cot\theta$, $\sec\theta$, and $\csc\theta$ of some frequently used angles.

θ(deg)	0°	30°	45°	60°	90°	120°	135°	150°	180°
θ(rad)	0	$\pi/6$	$\pi/4$	$\pi/3$	$\pi/2$	$2\pi/3$	$3\pi/4$	$5\pi/6$	π
$\tan\theta$	0	$\sqrt{3}/3$	1	$\sqrt{3}$	undefined	$-\sqrt{3}$	-1	$-\sqrt{3}/3$	0
$\cot\theta$	undefined	$\sqrt{3}$	1	$\sqrt{3}/3$	0	$-\sqrt{3}/3$	-1	$-\sqrt{3}$	undefined
$\sec\theta$	1	$2\sqrt{3}/3$	$\sqrt{2}$	2	undefined	-2	$-\sqrt{2}$	$-2\sqrt{3}/3$	-1
$\csc\theta$	undefined	2	$\sqrt{2}$	$2\sqrt{3}/3$	1	$2\sqrt{3}/3$	$\sqrt{2}$	2	undefined

θ(deg)	210°	225°	240°	270°	300°	315°	330°	360°
θ(rad)	$7\pi/6$	$5\pi/4$	$4\pi/3$	$3\pi/2$	$5\pi/3$	$7\pi/4$	$11\pi/6$	2π
$\tan\theta$	$\sqrt{3}/3$	1	$\sqrt{3}$	undefined	$-\sqrt{3}$	-1	$-\sqrt{3}/3$	0
$\cot\theta$	$\sqrt{3}$	1	$\sqrt{3}/3$	0	$-\sqrt{3}/3$	-1	$-\sqrt{3}$	undefined
$\sec\theta$	$-2\sqrt{3}/3$	$-\sqrt{2}$	-2	undefined	2	$\sqrt{2}$	$2\sqrt{3}/2$	1
$\csc\theta$	-2	$-\sqrt{2}$	$-2\sqrt{3}/3$	-1	$-2\sqrt{3}/3$	$-\sqrt{2}$	-2	undefined

Table 2

EXAMPLE 2. If $\tan \theta = -\frac{8}{15}$ and $\pi/2 < \theta < \pi$, find the values of the other trigonometric functions of θ.

Solution. First we note that, since $\tan \theta = -\frac{8}{15}$ and θ is an angle with its terminal side in the second quadrant, then $x = -15$, $y = 8$, and

$$r = \sqrt{(-15)^2 + (8)^2} = \sqrt{289} = 17$$

(see Figure 1). Therefore, we have the following values.

$$\sin \theta = \frac{8}{17} \qquad \csc \theta = \frac{17}{8} \qquad \cos \theta = -\frac{15}{17} \qquad \sec \theta = -\frac{17}{15} \qquad \cot \theta = -\frac{15}{8}$$

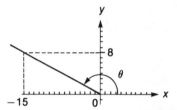

Figure 1

Self-test. If $\tan \theta = \frac{5}{12}$ and $\pi < \theta < 3\pi/2$, find each of the following.

(a) $\cot \theta =$ _____

(b) $x =$ _____, $y =$ _____, $r =$ _____

(c) $\sec \theta =$ _____

(d) $\sin \theta =$ _____

Next we consider graphing of the tangent, cotangent, secant, and cosecant functions. As we have indicated earlier, the tangent function is periodic with period π. Thus, the graph of $y = \tan \theta$ repeats the basic shape obtained on any interval with length of π. We note that $\tan 0 = 0$ and that as θ varies from 0 to $\pi/2$, $\tan \theta$ increases in value without bound. Also, as θ varies from 0 to $-\pi/2$, $\tan \theta$ decreases in value without bound. Plotting a few points that belong to the graph of $y = \tan \theta$, we obtain the portion of the graph on $[-\pi/2, \pi/2]$. It is usually helpful to draw the vertical lines $x = \pm\pi/2$, $\pm 3\pi/2$, $\pm 5\pi/2$, etc., which act as guides to the curve. The graph of $y = \tan \theta$ is shown in Figure 2.

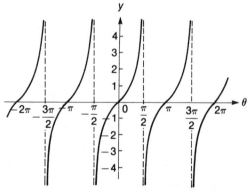

Figure 2

ANSWERS:

(a) $\frac{12}{5}$ (b) $x = -12$, $y = -5$, $r = 13$ (c) $-\frac{13}{12}$ (d) $-\frac{5}{13}$

Graphing the function $y = \cot \theta$ is similar to graphing the tangent function. We note that the cotangent is also periodic with period π, and that $\cot \pi/2 = 0$. Now, as θ varies from $\pi/2$ to π, $\cot \theta$ decreases in value without bound. Also, as θ varies from $\pi/2$ to 0, $\cot \theta$ increases without bound.

Locating a few points, such as $(\pi/4, 1)$ and $(3\pi/4, -1)$ that belong to the graph of $y = \cot \theta$, we plot the graph on $(0, \pi)$ and extend it to the right beyond π and to the left beyond 0, as shown in Figure 3.

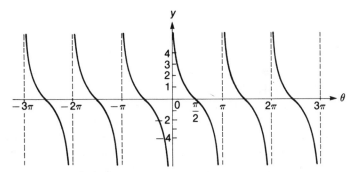

Figure 3

To graph the secant function, we recall that it is a periodic function with period 2π. We note that $\sec 0 = 1$ and that as θ varies from π to $\pi/2$, $\sec \theta$ increases without bound. Also, $\sec \pi = -1$ and, as θ varies from π to $\pi/2$, $\sec \theta$ decreases without bound. As θ varies from π to $3\pi/2$, $\sec \theta$ again decreases without bound. Finally, $\sec 2\pi = 1$ and as θ varies from 2π to $3\pi/2$, $\sec \theta$ increases without bound. Thus, we obtain the portion of the graph of $y = \sec \theta$ on $[0, 2\pi]$. We draw the vertical lines $x = \pm \pi/2$, $\pm 3\pi/2$, $\pm 5\pi/2$, etc., as guides to the curve. The graph of $y = \sec \theta$ (the solid line) is shown in Figure 4, together with the graph of $y = \cos \theta$.

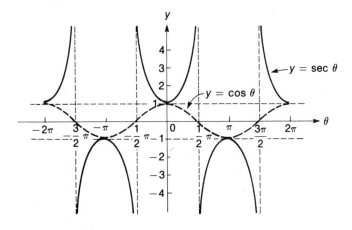

Figure 4

The graph of $y = \csc \theta$, which may be contrasted to the graph of the sine function, is similar to the graph of the secant function. Graphing the cosecant function is left as an exercise.

So far we have evaluated the trigonometric functions of some special angles. How do we find the value of a trigonometric function for any angle, say 47°10′? A table of values of trigonometric functions is included in Appendix B. These tables are constructed using electronic calculators and special formulas studied in higher mathematics. Two such formulas are

$$\sin x = x - x^3/3! + x^5/5! - x^7/7! + \ldots$$
$$\cos x = 1 - x^2/2! + x^4/4! - x^6/6! + \ldots$$

where x is any real number.

In Table 1 there are columns with headings for degrees, radians, and the six trigonometric functions. These columns give the values of the trigonometric functions to four decimal places for angles from 0° to 90° at intervals of 10 minutes. The first two columns give the measures of angles from 0° (or 0 radians) to 45° (or 0.7854 radians) increasing as we move downward. The last two columns give the measure of angles from 45° to 90° (or 1.5708 radians) increasing as we move upward. For angles in the first two columns, we use the trigonometric functions located at the top of the page, while for angles in the last two columns we use the trigonometric functions located at the bottom of the page.

These tables are used in one of two ways.

1.   Given an angle, we find the corresponding value of the function.
2.   Given the value of a function, we find the corresponding angle.

EXAMPLE 3. From Table 1 in Appendix B, find each of the following.

(a) sin 39°20′   (b) tan(−67°30′)   (c) cos 122°40′   (d) sec 43°47′

Solution.

(a)   In the column headed *Degrees* in Table 1 in Appendix B, we locate the entry 39°20′. Opposite this entry, in the column headed sin θ, we find the number 0.6338. Therefore, sin 39°20′ = 0.6338.

(b)   When we wish to find the function value of an angle less than 0°, we use an identity that would give us an equivalent function of an angle between 0° and 90°. In this case we note that $\tan(-\theta) = -\tan \theta$.

$$\tan(-67°30′) = -\tan 67°30′ = -2.414$$

(c)   When we wish to find the function value of an angle greater than 90°, we use an identity that would give us an equivalent function of an angle between 0° and 90°. In this case, we note that $\cos(\pi - \theta) = -\cos \theta$. Thus,

$$\cos 122°40′ = -\cos(180° - 122°40′) = -\cos 57°20′ = -0.5398$$

(d) When we wish to find a number which is not in the Tables, but lies between two entries of a Table, we use the method of **interpolation**. The angle $43°47'$ lies between the successive entries $43°40'$ and $43°50'$. We find the values of sec $43°40'$ and sec $43°50'$ and set up the following arrangement.

$$10'\left[\begin{array}{c} 7'\left[\begin{array}{l} \text{sec } 43°40'=1.382 \\ \text{sec } 43°47'=y \end{array}\right]d \\ \text{sec } 43°50'=1.386 \end{array}\right]0.004$$

The numbers on the side represent the difference between the indicated pairs of numbers. We need to find y. The difference between y and 1.382 is approximated by d and is assumed to be proportional to the change in θ. Thus we get

$$d/0.004 = {}^{7}\!/_{10} \quad \text{or} \quad d = {}^{7}\!/_{10}(0.004) \approx 0.003$$

Therefore,

$$y = \text{sec } 43°47' = 1.382 + 0.003 = 1.385$$

Self-test. Use Table 1 to fill in the blanks. Use an electronic calculator to check your answers.

(a) $\cos 8°40' = $ _____

(b) $\cot(-71°10') = $ _____

(c) $\sin 41°13' = $ _____

EXERCISES 2.5

In Exercises 1-4, use the definitions of the trigonometric functions to prove each identity. State the restrictions on θ or the function values of θ.

1. $\cos \theta \sec \theta = 1$ 2. $1 + \cot^2\theta = \csc^2\theta$
3. $\csc^2\theta(1 - \cos^2\theta) = 1$ 4. $\tan \theta \sec \theta(\csc \theta - \sin \theta) = 1$

In Exercises 5-20, find the values of the remaining five trigonometric functions of the angle satisfying the given conditions.

5. $\tan \theta = {}^{8}\!/_{15}, \pi < \theta < 3\pi/2$ 6. $\sin \theta = -{}^{8}\!/_{17}, 3\pi/2 < \theta < 2\pi$

7. $\cos \theta = {}^{5}\!/_{13}, 0 < \theta < \pi/2$ 8. $\cos \theta = -{}^{4}\!/_{5}, \pi < \theta < 3\pi/2$

ANSWERS:

9. $\cot \theta = -\frac{5}{3}$, $3\pi/2 < \theta < 2\pi$ 10. $\tan \theta = -\frac{5}{12}$, $\pi/2 < \theta < \pi$

11. $\cot \theta = \frac{3}{4}$, $\pi < \theta < 3\pi/2$ 12. $\sin \theta = \frac{2}{5}$, $\pi/2 < \theta < \pi$

13. $\sin \theta = \frac{2}{3}$, $\pi/2 < \theta < \pi$ 14. $\cos \theta = -\frac{1}{2}$, $\pi/2 < \theta < \pi$

15. $\tan \theta = 3$, $\pi < \theta < 3\pi/2$ 16. $\csc \theta = 3$, $\pi/2 < \theta < \pi$

17. $\sec \theta = 5$, $3\pi/2 < \theta < 2\pi$ 18. $\cot \theta = 5$, $\pi < \theta < 3\pi/2$

19. $\csc \theta = -\frac{17}{15}$, $3\pi/2 < \theta < 2\pi$ 20. $\cos \theta = -\frac{1}{3}$, $\pi/2 < \theta < \pi$

In Exercises 21-30, reduce each expression to one involving the sine and cosine functions only.

21. $(1 - \tan \theta)/(\cot \theta - 1)$ 22. $\cot \theta \csc \theta$

23. $\tan \theta \sec \theta$ 24. $\tan^2\theta + \cot^2\theta$

25. $\sec^2\theta - \csc^2\theta$ 26. $(\tan \theta - \sec \theta)/(\cot \theta - \csc \theta)$

27. $1 + \tan^2\theta$ 28. $\sec^2\theta + \csc^2\theta$

29. $\cot^2\theta - \tan^2\theta$ 30. $1 + \cot^2\theta$

In Exercises 31-46, sketch the graph of each function.

31. $y = \csc \theta$ 32. $y = \tan 2\theta$ 33. $y = \frac{1}{2} \sec \theta$

34. $y = 3 \tan \theta$ 35. $y = -\sec \theta$ 36. $y = -3 \sec \theta$

37. $y = -3 \csc \theta$ 38. $y = \csc 3\theta$ 39. $y = \sec 2\theta$

40. $y = \tan (\theta/2)$ 41. $y = \csc(\theta - \pi/4)$ 42. $y = \sec(\theta + \pi/3)$

43. $y = \sec(\theta + \pi/6)$ 44. $y = \cot(\theta - \pi/4)$ 45. $y = \tan(\theta + \pi/2)$

46. $y = -2 \tan \theta$

In Exercises 47-58, use Table 1 in Appendix B to find the sine, cosine, and tangent of each angle. Use an electronic calculator to check your answers.

47. $11°40'$ 48. $28°10'$ 49. $110°30'$

50. $-35°20'$ 51. $-125°50'$ 52. $275°10'$

53. $-316°$ 54. $-190°40'$ 55. $18°17'$

56. $117°36'$ 57. $-69°14'$ 58. $213°28'$

2.6 CIRCULAR FUNCTIONS AND APPLICATIONS

OBJECTIVES

1. *Define the circular functions in terms of coordinates of points on the unit circle.*
2. *Solve applied problems involving circular functions.*

We have defined the trigonometric functions as functions with a domain being the set of angle measures and a range being the set of all real numbers. The trigonometric functions are also defined as functions with a subset of the real numbers as its domain and a subset of real numbers as its range. In this section we shall see that this extension of the domain of the trigonometric functions to include real numbers is a matter of associating real numbers with the radian measure of the angles.

We will define the trigonometric functions of real numbers, called **circular functions** (or **wrapping functions**), in terms of the coordinates of points on the **unit circle**. We consider the unit circle with equation

$$x^2 + y^2 = 1$$

having its center at the origin and a radius of 1 (see Figure 1).

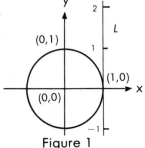

Figure 1

We construct a real number line L through the point $(1,0)$, parallel to the y-axis with 0 at the point $(1,0)$ and with a unit length the same as that of the coordinate system. Note that the circumference of the unit circle is of length 2π or approximately 6.28.

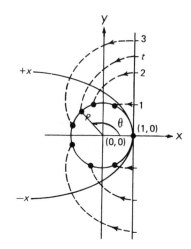

We now visualize the wrapping of L around the unit circle (see Figure 2). The upper ray above the point $(1,0)$, representing the positive real numbers, is wound in a counterclockwise direction. The lower ray, representing the negative real numbers, is wound in a clockwise direction. Each real number t is associated with a unique point on the graph of the unit circle as well as an angle θ in the standard position. We note, however, that each point on the unit circle represents infinitely many real numbers. Thus, if P represents a real number t, then it also represents every real number $t+2n\pi$ where n is an integer.

Figure 2

Now, the relationship between the number t and the angle θ is given by

$$\theta(\text{radians}) = \frac{t(\text{arc length})}{r(\text{radius})}$$

Since for the unit circle $r=1$, we have symbolically

$$\theta \ (\text{rad}) = t \qquad\qquad (1)$$

Using this association of angles in the standard position and real numbers, we may define the circular functions as follows (see Figure 3).

$\sin t = y = \sin \theta;$
$\csc t = 1/y = \csc \theta, \ y \neq 0;$
$\cos t = x = \cos \theta;$
$\sec t = 1/x = \sec \theta, \ x \neq 0;$
$\tan t = y/x = \tan \theta, \ x \neq 0;$
$\cot t = x/y = \cot \theta, \ y \neq 0;$

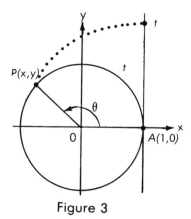

Figure 3

EXAMPLE 1. On the unit circle $x^2 + y^2 = 1$, find the point associated with each of the following real numbers.

(a) $\pi/4$.   (b) $-(\pi/6)$ (c) $8\pi/3$

Solution.

(a) Using Equation 1, the number $\pi/4$ is associated with the angle $\pi/4$ radians in the standard position. Thus, the coordinates of the point are $(\sqrt{2}/2, \sqrt{2}/2)$. See point P in Figure 4.

(b) The number $-(\pi/6)$ is associated with the angle $-(\pi/6)$ radians in the standard position. Thus, the coordinates of the point are $(\sqrt{3}/2, -\frac{1}{2})$. See point Q in Figure 4.

(c) In this case the point associated with the real number $8\pi/3$ has coordinates $(-\frac{1}{2}, \sqrt{3}/2)$. See point R in Figure 4.

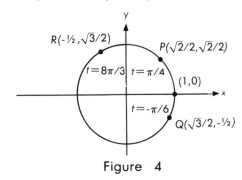

Figure 4

EXAMPLE 2. Find the circular function values for the real number t, if t corresponds to the point $(-\sqrt{3}/2, \frac{1}{2})$.

Solution. The angle associated with the point $(-\sqrt{3}/2, \frac{1}{2})$ is $5\pi/6$ radians. Hence, $t = 5\pi/6$ and we have

$$\sin t = \frac{1}{2} \qquad \csc t = 2 \qquad \cos t = -\sqrt{3}/2$$

$$\sec t = -2\sqrt{3}/3 \qquad \tan t = (\frac{1}{2})/-\sqrt{3}/2 = -\sqrt{3}/3 \qquad \cot t = -\sqrt{3}$$

Self-test.

(a) On the unit circle, the point associated with the real number $t = -\pi/2$ is _____, and with $t = 11\pi/3$ is _____.

(b) If the real number t corresponds to the point $(-1,0)$ then

$\sin t =$ _____, $\cos t =$ _____, $\tan t =$ _____

$\csc t =$ _____, $\sec t =$ _____, $\cot t =$ _____

It should be clear that the properties we derived for the trigonometric functions of angles also apply to the circular function. Also, the graphs of the circular functions are equivalent to the graphs of the corresponding trigonometric functions.

ANSWERS: (a) $(0,-1), (\frac{1}{2})''(1, -\sqrt{3}/2)$ (b) $0, -1, 0,$ undefined, $-1,$ undefined.

We now turn our attention to the applications of the circular functions. There are many natural phenomena in which a recurring pattern of behavior is exhibited. The motion of a clock pendulum, sound waves, the motion of the planets, certain ecological systems, water waves, light and other electromagnetic waves are examples. Many such phenomena can be described by an equation of the form

$$y = A \sin(Bt + C) \tag{2}$$

or

$$y = A \cos(Bt + C) \tag{3}$$

where $A, B,$ and C are constants. The student should recall that $|A|$ is the amplitude, $2\pi/B$ is the period, and $-C/B$ is the phase shift. Phenomena that can be explained by Equations 1 or 2 are said to be **simple harmonic**. If t represents time, then the reciprocal of the period, that is $B/2\pi$, is called the **frequency**. For example, if a given sine wave has a period of $\frac{1}{20}$th of a second, then one cycle is completed in $\frac{1}{20}$th of a second and the wave is said to have a frequency of 20 cycles per second (cps).

We now consider a system which involves simple harmonic motion. Suppose that an object is attached to a spring as shown in Figure 5.

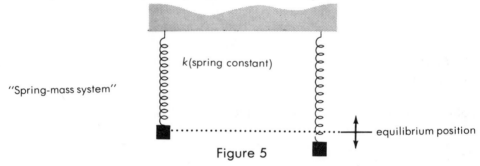

"Spring-mass system"

k(spring constant)

equilibrium position

Figure 5

When the object is at rest, we say that it is in equilibrium position. If the object is pulled down and then released, it will oscillate about the equilibrium position. Using physical laws and calculus, it can be shown that under certain conditions the motion of the object can be described mathematically by the equation

$$y = A \cos \omega t \tag{4}$$

where y is the displacement measured from the equilibrium point (positive upward), A and ω are constants dependent upon the particular spring-mass system, and t is a measure of time. The constant ω is called the **natural angular frequency** and is given by

$$\omega = \sqrt{k/m} \tag{5}$$

where k is the spring constant and m is the mass of the object attached to the spring. The constant A represents the initial displacement and sets the bounds about the equilibrium point through which the object oscillates. The frequency of oscillation is given by $\omega/2\pi$ cycles per unit of time.

EXAMPLE 3. Suppose Equation 4 describes the motion of a given spring-mass system in which $\omega = 4$. If the initial displacement is 10 cm, find the distance of the object from the equilibrium position when

(a) $t = \pi/2$ seconds   (b) $t = 1.1$ seconds

Solution.

(a) Since the inital displacement is 10 and $\omega=4$, then Equation 4 becomes

$$y=10 \cos 4t$$

and when $t=\pi/2$, we have

$$y=10 \cos 4(\pi/2)=10 \cos 2\pi=10$$

That is, at $t=\pi/2$ seconds, the object is 10 cm above the point of equilibrium.

(b) When $t=1.1$ seconds, we have

$$y=10 \cos 4(1.1)=10 \cos 4.4=10(-0.31)=-3.1$$

That is, when $t=1.1$ seconds, the object is 3.1 cm below the point of equilibrium.

EXAMPLE 4. Find the natural angular frequency of a spring-mass system with a mass of 4 and a spring constant of 0.7.

Solution. Using Equation 5 we have

$$\omega=\sqrt{k/m} = \sqrt{0.7/4}=0.42$$

Self-test. Consider a spring-mass system in which the spring constant is 0.8 and the mass is 3.

(a) Find the natural angular frequency: $\omega=$ _____

(b) Find the position of the mass at $t=2$ seconds if the initial displacement is 6 cm. $y=$ _____

Historically, the problem of the spring-mass system played an important part in the development of physics. Today, although the spring-mass system in itself does not attract much attention, it exhibits behavior of more complex systems which are being analyzed. Motion of the surface of the ocean and the motion of pendulum-like mechanisms are examples.

A simple sound wave is produced by a vibrating object such as a tuning fork, a violin, human vocal cords, drums, and so on. The vibrating object sets air molecules in motion which in turn cause a periodic change in the air pressure. The periodic change in air pressure travels through the air at about 330 meters per second. Upon reaching your eardrum, the periodic change in air pressure causes the drum to vibrate at the same frequency as the source. This is then transmitted to your brain in the form of sound.

A simple sound wave such as a tuning fork can be described by an equation of the form

$$y=A \sin 2\pi ft \tag{6}$$

where A is the amplitude, f is the frequency, and t is the time.

ANSWERS:

Self-test. If a sound wave is described by
$$y=0.06 \sin 800\pi t$$
Then
(a) The amplitude is _____.
(b) The period is _____.
(c) The frequency is _____.

Most sound waves are more complex than Equation 6 indicates. However, these complex sound waves can be described in terms of simple sine waves.

EXERCISES 2.6

In Exercises 1-12, find the point on the unit circle associated with the given real number.

1. $\pi/3$ 2. $-2\pi/3$ 3. $\pi/6$ 4. $-\pi/4$

5. $-(3\pi/4)$ 6. $7\pi/6$ 7. $3\pi/2$ 8. $-5\pi/2$

9. $5\pi/6$ 10. $-13\pi/2$ 11. $10\pi/3$ 12. $19\pi/6$

In Exercises 13-21, use Table 1 to find the circular function values of the given number.

13. 0.34   14. 1.17   15. -1.05 16. -1.12

17. -0.71 18. $-\pi/2$ 19. $-3\pi/2$ 20. $\pi/4$

21. $3\pi/4$

ANSWERS:

(c) 400   (b) 1/400   (a) 0.06

In Exercises 22-35, find the six circular function values for the real number t, if t corresponds to the given point.

22. $(\sqrt{3}/2, \frac{1}{2})$

23. $(\sqrt{3}/2, -\frac{1}{2})$

24. $(1,0)$

25. $(0,1)$

26. $(0,-1)$

27. $(\sqrt{2}/2, -\sqrt{2}/2)$

28. $(-\sqrt{2}/2, \sqrt{2}/2)$

29. $(-\sqrt{3}/2, -\frac{1}{2})$

30. $(\sqrt{10}/10, 3\sqrt{10}/10)$

31. $(-\sqrt{10}/10, 3\sqrt{10}/10)$

32. $(1/\sqrt{5}, 2/\sqrt{5})$

33. $(2/\sqrt{5}, -1/\sqrt{5})$

34. $(-\frac{5}{13}, -\frac{12}{13})$

35. $(3/\sqrt{13}, -2/\sqrt{13})$

In Exercises 36-38, use the definitions of the circular functions to verify the given identities.

36. $\sin^2 t + \cos^2 t = 1$

37. $1 + \tan^2 t = \sec^2 t$

38. $1 + \cot^2 t = \csc^2 t$

In Exercises 39-42, find the value of the $\sec t$, $\csc t$, and $\tan t$ if

39. $\cos t = -\frac{3}{5}$, $\sin t = \frac{4}{5}$

40. $\sin t = \frac{15}{17}$, $\cos t = -\frac{8}{17}$

41. $\sin t = -\sqrt{5}/5$, $\cos t = -2/\sqrt{5}$

42. $\sin t = -\sqrt{2}/2$, $\cos t = \sqrt{2}/2$

In Exercises 43-44, a spring-mass system is considered with the equation of motion $y = A \cos \omega t$.

43. If the initial displacement is 8 centimeters and $\omega = 5$, find y when $t = 2$ and when $t = 4$.

44. If the initial displacement is -6, and $\omega = 5$, find y when $t = 4$ and $t = 6$.

45. Find the natural angular frequency of a spring-mass system with a mass of 8 and a spring constant of 0.9.

46. If a sound wave is described by

$$y = 0.006 \sin 1{,}000\pi t$$

find its amplitude, period, and frequency if t is measured in seconds.

47. The strength of the current I in an alternating current system is given by

$$I = 0.05 \sin 120\pi t$$

where I is measured in amperes and it is expressed in seconds. Find the current at

(a) $t = 0.01$ (b) $t = 0.2$ (c) $t = 0.4$ (d) $t = 1$

48. Let $I=A \sin \omega t$ describe the current in an alternating current generator. If the generator produces a current flow of 60 cycles per second and a maximum current value of 15 amperes, write an equation describing the current flow.

49. Suppose an alternating current generator produces a current described by the equation

$$I=5 \sin(377t-60\pi)$$

where I is the current and t is measured in seconds. Find the amplitude, period, frequency, and phase shift for the current.

50. A water wave has an equation of the form

$$y=25 \sin 0.3\pi t$$

(a) Find the height of the wave from trough to crest.

(b) Find the wavelength.

51. Archimedes' principle states that an object submerged in a liquid is buoyed up by a force equal to the weight of the liquid displaced. With the aid of calculus, it is found that the equation of the up and down motion of the object is given by

$$y=B \cos \sqrt{(\varrho A/m)} \; t$$

where B is the initial displacement, m is the mass of the object, A is its cross-sectional area, ϱ is the density of the liquid, and t is the time.

   Suppose a cubic block of wood weighing 96 pounds has a cross-sectional area of 4 square feet. If it is depressed in a liquid which weighs 50 pounds per cubic foot and released, what is the period of oscillation? (Note, to find the mass of the block, use the equation $m=\text{weight}/\text{acceleration}$ where the acceleration is 32 ft/sec^2).

52. If a cubic block of wood 2 feet long on one side is observed to bob up and down in water with a period of 1 second, how much does it weigh? (Note, period $=2\pi\sqrt{\varrho A/m}$).

REVIEW OF CHAPTER TWO

1. What is the vertex of an angle?

2. What is an angle in standard position?

3. Define the degree measure of an angle.

4. Define the radian measure of an angle.

5. Fill in the blanks:
   - (a) $1° =$ _____ radians.
   - (b) 1 radian = _____ degrees.
   - (c) $1° =$ _____ minutes = _____ seconds.

6. If an angles measures $26°34'20''$, what is its radian measure?

7. What is the degree measure equivalent to 1.36 radians?

8. What are coterminal angles?

9. If α and β are coterminal, with $\alpha = 40°$ and $-2\pi \geq \beta > 0$, what is the measure of β?

10. Define angular speed.

11. Define linear speed.

12. A wheel has a diameter of 4 feet and makes 3 revolutions per second. Find the linear speed in ft/min of a point on its rim.

13. Define the six trigonometric functions and give the domain and range of each.

14. Sketch a graph of each of the six trigonometric functions over one period.

15. Find the sine, cosine, and tangent of

    (a) $2\pi/3$ (b) $-\pi/4$ (c) $-5\pi/2$

16. Let θ be an angle in standard position with its terminal side containing the point $(3, -4)$. Find the values of the six trigonometric functions of θ.

17. Define a periodic function.

18. What is the period of each of the six trigonometric functions?

19. State the Pythagorean identity.

20. Prove that $\sin(-\theta)=-\sin\theta$ and $\cos(-\theta)=\cos\theta$.

21. Fill in the blanks.

$$\sin(\theta+2n\pi)= \underline{\hspace{6cm}}, n\in Z$$
$$\cos(\theta+2n\pi)= \underline{\hspace{6cm}}, n\in Z$$

22. Indicate the sign of the six trigonometric functions in the four quadrants of a rectangular coordinate system.

23. Define amplitude.

24. Define cycle.

25. Define phase shift.

26. For the function $y=A\cos(B\theta+C)$, give:
    (a) The amplitude.
    (b) The period.
    (c) The phase shift.

27. Reduce the following expression to one involving the sine and cosine functions only.

$$(\tan\theta-\sec\theta)/(\cot\theta-\csc\theta)$$

28. Define the circular functions.

29. In the equation $y=A\sin(Bt+C)$ what is the frequency?

TEST TWO

1. Convert to radian measure each angle of the following measures. Indicate the quadrant in which the terminal side lies.

    (a) 32°   (b) −135°   (c) 167°

2. Convert to degree measure.

    (a) 5π/2 radians   (b) 3π radians

3. Find the radian measure of a central angle subtended by an arc of 18 cm on a circle with a radius of 8 cm.

4. A car is traveling at 45 kilometers per hour. If the diameter of the wheel is 66 cm, what is the angular speed in radians/sec of a point on the edge?

5. Find the values of the six trigonometric functions of a positive angle in standard position whose terminal side contains the point $(-2,3)$.

6. Without using tables, compute the following:

    (a) cos(−315°)   (b) tan 495°   (c) sin 210°

7. Given that $\cos \theta = \frac{4}{5}$ and $\sin \theta < 0$, find the other five function values of θ.

8. Draw a graph of $y = 2 \cos(\theta + \pi/4)$ on $[-2\pi, 2\pi]$. Indicate the amplitude, period, and phase shift.

9. Find the six circular function values for the real number t, if t corresponds to the point $(\frac{1}{2}, -\sqrt{3}/2)$.

10. Using the definition of the circular functions, prove that $1 + \cot^2 t = \csc^2 t$.

11. Given that $\cos \theta = 0.3090$, find $\sin(\pi/2 - \theta)$.

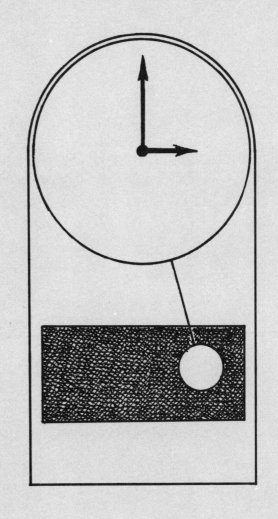

CHAPTER THREE:
ANALYTIC
TRIGONOMETRY

3.1 THE ADDITION FORMULAS

OBJECTIVES

1. *Establish formulas for the sine, cosine and tangent of the sum of two angles.*
2. *Use these formulas to prove certain identities.*

It is often useful to change a trigonometric expression from one form to an equivalent form. This involves the use of identities. Equations that involve trigonometric function values are called **trigonometric equations**. A trigonometric equation that is true for all permissible values of the variables involved is called a **trigonometric identity**. In Chapter 2, we considered the following identities.

$$\csc \theta = 1/\sin \theta \qquad \sec \theta = 1/\cos \theta \qquad \tan \theta = \sin \theta/\cos \theta$$

$$\cot \theta = \cos \theta/\sin \theta \qquad \sin^2\theta + \cos^2\theta = 1 \qquad \sin(-\theta) = -\sin \theta$$

$$\cos(-\theta) = \cos \theta \qquad \tan(-\theta) = -\tan \theta \qquad 1 + \tan^2\theta = \sec^2\theta$$

$$1 + \cot^2\theta = \csc^2\theta$$

There are a large number of identities connecting the various trigonometric functions. First we shall consider the sine and cosine of the sum of two angles. We note that, in general

$$\sin(\theta_1 + \theta_2) \neq \sin \theta_1 + \sin \theta_2$$

For example, take $\theta_1 = \pi/3$ and $\theta_2 = \pi/6$. Then,

$$\sin(\theta_1 + \theta_2) = \sin(\pi/3 + \pi/6) = \sin(\pi/2) = 1$$

whereas,

$$\sin \theta_1 + \sin \theta_2 = \sin(\pi/3) + \sin(\pi/6)$$
$$= \sqrt{3}/2 + \tfrac{1}{2}$$
$$= (\sqrt{3}+1)/2 \approx 1.37$$

Thus,

$$\sin(\pi/3 + \pi/6) \neq \sin(\pi/3) + \sin(\pi/6)$$

Likewise, we can show that in general

$$\cos(\theta_1 + \theta_2) \neq \cos \theta_1 + \cos \theta_2$$

Self-test.

(a) $\cos(\pi/3 + \pi/6) = \cos$ _____ = _____.

(b) $\cos(\pi/3) =$ ____; $\cos(\pi/6) =$ ____, and $\cos(\pi/3) + \cos(\pi/6) =$ ____.

(c) Is $\cos(\pi/3 + \pi/6) = \cos(\pi/3) + \cos(\pi/6)$? _____

The relation between $f(\theta_1 + \theta_2)$ and $f(\theta_1)$ and $f(\theta_2)$, where f is a trigonometric function will be established in this section.

We consider the cosine function.

THEOREM 3.1 (Addition Formulas for Cosine). If θ_1 and θ_2 are any two angles, then

$$\cos(\theta_1 - \theta_2) = \cos \theta_1 \cos \theta_2 + \sin \theta_1 \sin \theta_2 \qquad (1)$$

and

$$\cos(\theta_1 + \theta_2) = \cos \theta_1 \cos \theta_2 - \sin \theta_1 \sin \theta_2 \qquad (2)$$

Proof. Let θ_1 and θ_2 be any two angles measured in radians or degrees and placed in standard position (see Figure 1). Place the angle $\theta_1 - \theta_2$ in standard position. With the center at the origin, draw a unit circle. Let $B, C,$ and D be the points of intersection of the circle with the terminal sides of $\theta_1 - \theta_2$, θ_2 and θ_1, respectively. Since $\cos \theta_1 = x/1$, $\sin \theta_1 = y/1$ and so on, we obtain the coordinates of $B, C,$ and D:

$$B = [\cos(\theta_1 - \theta_2), \ \sin(\theta_1 - \theta_2)]$$
$$C = (\cos \theta_2, \ \sin \theta_2)$$
$$D = (\cos \theta_1, \ \sin \theta_1)$$

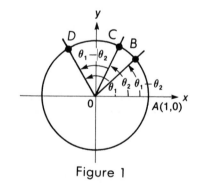

Figure 1

ANSWERS:

From geometry, we recall that

$$[d(A,B)]^2 = [d(D,C)]^2$$

since the line segment from A to B is subtended by an angle whose measure is the same as the angle subtending the line segment from D to C. Now, using the distance formula, we have:

$$
\begin{aligned}
[d(A,B)]^2 &= [1-\cos(\theta_1-\theta_2)]^2+[0-\sin(\theta_1-\theta_2)]^2 \\
&= 1-2\cos(\theta_1-\theta_2)+\cos^2(\theta_1-\theta_2)+\sin^2(\theta_1-\theta_2) \\
&= 2-2\cos(\theta_1-\theta_2) \tag{3}
\end{aligned}
$$

Also

$$
\begin{aligned}
[d(D,C)]^2 &= (\cos\theta_1-\cos\theta_2)^2+(\sin\theta_1-\sin\theta_2)^2 \\
&= \cos^2\theta_1-2\cos\theta_1\cos\theta_2+\cos^2\theta_2+\sin^2\theta_1-2\sin\theta_1\sin\theta_2+\sin^2\theta_2 \\
&= (\cos^2\theta_1+\sin^2\theta_1)+(\cos^2\theta_2+\sin^2\theta_2)-2(\cos\theta_1\cos\theta_2+\sin\theta_1\sin\theta_2) \\
&= 2-2(\cos\theta_1\cos\theta_2+\sin\theta_1\sin\theta_2) \tag{4}
\end{aligned}
$$

Equating Equations 3 and 4, we get

$$2-2\cos(\theta_1-\theta_2)=2-2(\cos\theta_1\cos\theta_2+\sin\theta_1\sin\theta_2)$$

Simplifying, we obtain Equation 1:

$$\cos(\theta_1-\theta_2)=\cos\theta_1\cos\theta_2+\sin\theta_1\sin\theta_2$$

To prove Equation 2, we let $\theta_2=-\alpha$. Then

$$
\begin{aligned}
\cos(\theta_1+\theta_2) &= \cos(\theta_1-\alpha) \\
&= \cos\theta_1\cos\alpha+\sin\theta_1\sin\alpha \\
&= \cos\theta_1\cos(-\theta_2)+\sin\theta_1\sin(-\theta_2) \\
&= \cos\theta_1\cos\theta_2-\sin\theta_1\sin\theta_2
\end{aligned}
$$

This proves Theorem 3.1.

For the sine function we have the following result.

THEOREM 3.2   (Addition Formulas for Sine) If θ_1 and θ_2 are any two angles, then

$$\sin(\theta_1-\theta_2)=\sin\theta_1\cos\theta_2-\sin\theta_2\cos\theta_1 \tag{5}$$

and

$$\sin(\theta_1+\theta_2)=\sin\theta_1\cos\theta_2+\sin\theta_2\cos\theta_1 \tag{6}$$

Proof. From Theorem 3.1,

$$\cos(\pi/2-\theta) = \cos(\pi/2)\cos\theta + \sin(\pi/2)\sin\theta$$
$$= \sin\theta$$

Thus,

$$\sin(\theta_1-\theta_2) = \cos[\pi/2-(\theta_1-\theta_2)]$$
$$= \cos[(\pi/2-\theta_1)+\theta_2]$$
$$= \cos(\pi/2-\theta_1)\cos\theta_2 - \sin(\pi/2-\theta_1)\sin\theta_2$$
$$= \sin\theta_1\cos\theta_2 - \cos\theta_1\sin\theta_2$$

where $\cos(\pi/2-\theta_1)=\sin\theta_1$ and $\sin(\pi/2-\theta_1)=\cos\theta_1$. (See Exercises 41 and 42 in Section 2.3). This establishes Formula 5. To establish Formula 6, let $\theta_2=-\alpha$. Then

$$\sin(\theta_1+\theta_2) = \sin(\theta_1-\alpha)$$
$$= \sin\theta_1\cos\alpha - \sin\alpha\cos\theta_1$$
$$= \sin\theta_1\cos(-\theta_2) - \sin(-\theta_2)\cos\theta_1$$
$$= \sin\theta_1\cos\theta_2 + \sin\theta_2\cos\theta_1$$

This proves Theorem 3.2.

Using Theorems 3.1 and 3.2 we obtain the following results for the tangent function.

THEOREM 3.3   (Addition Formulas for the Tangent) If θ_1 and θ_2 are two angles, then,

$$\tan(\theta_1-\theta_2) = \frac{\tan\theta_1-\tan\theta_2}{1+\tan\theta_1\tan\theta_2} \tag{7}$$

$$\tan(\theta_1+\theta_2) = \frac{\tan\theta_1+\tan\theta_2}{1-\tan\theta_1\tan\theta_2} \tag{8}$$

Proof. We have,

$$\tan(\theta_1-\theta_2) = \frac{\sin(\theta_1-\theta_2)}{\cos(\theta_1-\theta_2)}$$

$$= \frac{\sin\theta_1\cos\theta_2-\sin\theta_2\cos\theta_1}{\cos\theta_1\cos\theta_2+\sin\theta_1\sin\theta_2}$$

Dividing the numerator and the denominator by $\cos\theta_1\cos\theta_2$ we obtain,

$$\tan(\theta_1-\theta_2) = \frac{\dfrac{\sin\theta_1\cos\theta_2}{\cos\theta_1\cos\theta_2}-\dfrac{\sin\theta_2\cos\theta_1}{\cos\theta_1\cos\theta_2}}{\dfrac{\cos\theta_1\cos\theta_2}{\cos\theta_1\cos\theta_2}+\dfrac{\sin\theta_1\sin\theta_2}{\cos\theta_1\cos\theta_2}} = \frac{\tan\theta_1-\tan\theta_2}{1+\tan\theta_1\tan\theta_2}$$

A similar procedure gives

$$\tan(\theta_1 + \theta_2) = \frac{\tan \theta_1 + \tan \theta_2}{1 - \tan \theta_1 \tan \theta_2}$$

This proves Theorem 3.3.

EXAMPLE 1. Find sin 15° using an addition formula.

Solution. We note that 15° = 60° − 45°. Using Formula 5, we obtain

$$\begin{aligned}
\sin 15° &= \sin(60° - 45°) \\
&= \sin 60° \cos 45° - \sin 45° \cos 60° \\
&= (\sqrt{3}/2)(\sqrt{2}/2) - (\sqrt{2}/2)(\tfrac{1}{2}) \\
&= (\sqrt{2}/4)(\sqrt{3} - 1)
\end{aligned}$$

EXAMPLE 2. Find sec 75°.

Solution. We note that 75° = 45° + 30° and that sec $\theta = (1/\cos \theta)$. Thus,

$$\begin{aligned}
\sec 75° &= 1/\cos 75° \\
&= 1/\cos(45° + 30°) \\
&= 1/(\cos 45° \cos 30° - \sin 45° \sin 30°) \\
&= 1/[(\sqrt{2}/2 \times \sqrt{3}/2) - (\sqrt{2}/2 \times \tfrac{1}{2})] \\
&= 1/(\sqrt{2}/4)(\sqrt{3} - 1) \\
&= 4/[\sqrt{2}(\sqrt{3} - 1)]
\end{aligned}$$

Self-test. Use the addition formulas to find each of the following:

(a) cos 15° = _____

(b) tan 15° = _____

EXAMPLE 3. Given that cos $\alpha = -\tfrac{4}{5}$ where α is in quadrant II and sin $\beta = \tfrac{12}{13}$ where β is in quadrant I, find

(a) $\cos(\alpha + \beta)$ (b) $\tan(\alpha - \beta)$

Solution. In Figure 2, the angles α and β are represented geometrically. Note that there is no loss of generality in considering α and β as positive angles between 0° and 360°.

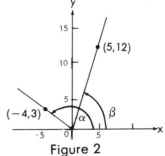

Figure 2

Since $\cos \alpha = -\frac{4}{5}$, the point $(-4,3)$ is on the terminal side of α. Similarly, since $\sin \beta = \frac{12}{13}$, the point $(5,12)$ is on the terminal side of β. Thus, we see that

$$\sin \alpha = \frac{3}{5}, \ \tan \alpha = -\frac{3}{4}, \ \cos \beta = \frac{5}{13}, \text{ and } \tan \beta = \frac{12}{5}$$

(a) Using Identity 2, we find

$$\begin{aligned}
\cos(\alpha + \beta) &= \cos \alpha \cos \beta - \sin \alpha \sin \beta \\
&= (-\tfrac{4}{5})(\tfrac{5}{13}) - (\tfrac{3}{5})(\tfrac{12}{13}) \\
&= -\tfrac{56}{65}
\end{aligned}$$

(b) Using Formula 7, we find

$$\tan(\alpha - \beta) = \frac{\tan \alpha - \tan \beta}{1 + \tan \alpha \tan \beta}$$

$$= \frac{(-\tfrac{3}{4}) - (\tfrac{12}{5})}{1 + (-\tfrac{3}{4})(\tfrac{12}{5})}$$

$$= \tfrac{63}{16}$$

Self-test. Given that $\sin \alpha = \frac{3}{5}$, where α is an angle in quadrant I, and $\cos \beta = -\frac{5}{13}$, where β is an angle in quadrant II, find the following:

(a) $\cos \alpha =$ _____, (b) $\sin \beta =$ _____

(c) $\sin(\alpha + \beta) =$ _____, (d) $\cos(\alpha - \beta) =$ _____

EXAMPLE 4. Prove that

$$\sin(\alpha + \beta) + \sin(\alpha - \beta) = 2 \sin \alpha \cos \beta$$

Solution. Using Formulas 5 and 6, we find

$$\sin(\alpha + \beta) + \sin(\alpha - \beta) = \sin \alpha \cos \beta + \sin \beta \cos \alpha + \sin \alpha \cos \beta - \sin \beta \cos \alpha$$
$$= 2 \sin \alpha \cos \beta$$

The next example illustrates an important application of an addition formula.

EXAMPLE 5. Show that for every θ

$$a \cos B\theta + b \sin B\theta = A \cos(B\theta - C) \tag{9}$$

where, $A = \sqrt{a^2 + b^2}$ and $\tan C = b/a$.

Solution. Using the addition Formula 1, we have

$$A \cos(B\theta - C) = A(\cos B\theta \cos C + \sin B\theta \sin C)$$

Now the equality

$$A \cos B\theta \cos C + A \sin B\theta \sin C = a \cos B\theta + b \sin B\theta$$

holds true for every θ if and only if

$$a = A \cos C \quad \text{and} \quad b = A \sin C \tag{10}$$

Consequently, from the relations in 10 we get

$$a^2 + b^2 = A^2\cos^2C + A^2\sin^2C = A^2(\cos^2C + \sin^2C) = A^2$$

Thus, we may choose $A = \sqrt{a^2 + b^2}$. Also from 10 we note that

$$b/a = (A \sin C)/(A \cos C) = (\sin C)/(\cos C) = \tan C$$

This establishes Formula 9.

EXERCISES 3.1

In Exercises 1-18, use the addition formulas to compute the value of each expression.

1. $\sin 75°$ 2. $\tan 75°$ 3. $\cos 75°$ 4. $\csc 15°$

5. $\cot 15°$ 6. $\sec 15°$ 7. $\sin 195°$ 8. $\cos 165°$

9. $\tan 165°$ 10. $\sin 255°$ 11. $\cos 375°$ 12. $\tan 345°$

13. $\sin 285°$ 14. $\cos 435°$ 15. $\cos(-\pi/12)$ 16. $\tan (5\pi/12)$

17. $\cos(5\pi/12)$ 18. $\sin(7\pi/12)$

19. Prove Formula 8.

In Exercises 20-25, evaluate the given expression if $\sin \alpha = \frac{3}{5}$ and $\cos \beta = \frac{12}{13}$, with $0 < \alpha < \pi/2$ and $0 < \beta < \pi/2$.

20. $\sin(\alpha + \beta)$ 21. $\cos(\alpha - \beta)$ 22. $\cos(\alpha + \beta)$ 23. $\sin(\alpha - \beta)$

24. $\tan(\alpha - \beta)$ 25. $\tan(\alpha + \beta)$

In Exercises 26-31, evaluate the given expression if $\tan \alpha = -\frac{8}{15}$ and $\cos \beta = -\frac{4}{5}$, with $\pi/2 < \alpha < \pi$ and $\pi < \beta < 3\pi/2$.

26. $\cos(\alpha - \beta)$ 27. $\sin(\alpha + \beta)$ 28. $\sin(\alpha - \beta)$ 29. $\cos(\alpha + \beta)$

30. $\tan(\alpha + \beta)$ 31. $\tan(\alpha - \beta)$

In Exercises 32-46, prove each identity.

32. $\cos(\alpha + \beta) + \cos(\alpha - \beta) = 2 \cos \alpha \cos \beta$

33. $\sin(\alpha + \beta) - \sin(\alpha - \beta) = 2 \cos \alpha \sin \beta$

34. $\cos(\alpha - \beta) - \cos(\alpha + \beta) = 2 \sin \alpha \sin \beta$

35. $\sin(\alpha + \beta)\sin(\alpha - \beta) = \sin^2\alpha - \sin^2\beta$

36. $\cos(\alpha + \beta)\cos(\alpha - \beta) = \cos^2\alpha - \sin^2\beta$

37. $\cos(\alpha + 3\pi/2) = \sin \alpha$

38. $\sin(\theta - 3\pi/2) = \cos \theta$

39. $\cos(\theta + \pi/4) = (\sqrt{2}/2)(\cos \theta - \sin \theta)$

40. $\sin(\theta + \pi/4) = (\sqrt{2}/2)(\sin \theta + \cos \theta)$

41. $\tan(\theta + \pi/4) = (1 + \tan \theta)/(1 - \tan \theta)$

42. $\cot(\theta - \pi/3) = (\sqrt{3} \tan \theta - 1)/(\tan \theta - \sqrt{3})$

43. $\sin \alpha + \sin \beta = 2 \sin[(\alpha + \beta)/2] \cos[(\alpha - \beta)/2]$

44. $\sin \alpha - \sin \beta = 2 \cos[(\alpha + \beta)/2] \sin[(\alpha - \beta)/2]$

45. $\cos \alpha + \cos \beta = 2 \cos[(\alpha + \beta)/2] \cos[(\alpha - \beta)/2]$

46. $\cos \alpha - \cos \beta = -2 \sin[(\alpha + \beta)/2] \sin[(\alpha - \beta)/2]$

In Exercises 47-49, write the given sums as products.

47. $\cos 3\theta + \cos \theta$ 48. $\sin 5\theta + \sin \theta$

49. $\sin[2\alpha/(\alpha^2 - \beta^2)] + \sin[2\beta/(\alpha^2 - \beta^2)]$

50.   If α and β are complementary angles, show that
$$\sin^2\alpha + \sin^2\beta = 1$$

3.2 DOUBLE-ANGLE AND
HALF-ANGLE FORMULAS

OBJECTIVES

1. Establish formulas for the trigonometric functions of 2θ and $\theta/2$.
2. Use these formulas to prove certain identities.

Formulas for trigonometric functions of 2θ in terms of θ can be derived from formulas derived in the previous section. For example

$$\sin 2\theta = \sin(\theta+\theta) = \sin \theta\cos \theta + \cos \theta\sin \theta = 2 \sin \theta\cos \theta \tag{1}$$

Similarly, the following **double-angle formulas** can be readily established.

$$\cos 2\theta = \cos^2\theta - \sin^2\theta \tag{2}$$
$$\cos 2\theta = 1 - 2 \sin^2\theta \tag{3}$$
$$\cos 2\theta = 2 \cos^2\theta - 1 \tag{4}$$
$$\tan 2\theta = (2 \tan \theta)/(1 - \tan^2\theta) \tag{5}$$

EXAMPLE 1. Prove Formula 4.

Solution. Use the addition formula for cosine and the Pythagorean identity to obtain

$$\begin{aligned}
\cos 2\theta &= \cos(\theta+\theta) \\
&= \cos \theta\cos \theta - \sin \theta\sin \theta \\
&= \cos^2\theta - \sin^2\theta \\
&= \cos^2\theta - (1 - \cos^2\theta) \\
&= 2 \cos^2\theta - 1
\end{aligned}$$

EXAMPLE 2. Express $\cos 4\theta$ in terms of functions of θ.

Solution.

$$\cos 4\theta = \cos 2(2\theta)$$
$$= 2 \cos^2 2\theta - 1$$
$$= 2(2 \cos^2\theta - 1)^2 - 1$$
$$= 2(4 \cos^4\theta - 4 \cos^2\theta + 1) - 1$$
$$= 8 \cos^4\theta - 8 \cos^2\theta + 1$$

EXAMPLE 3. Find $\sin 2\theta$ if $\sin \theta = -\frac{1}{4}$ and θ is in quadrant IV.

Solution. In Figure 1, we show θ in the standard position. We note that $\cos \theta = \sqrt{15}/4$ and hence,

$$\sin 2\theta = 2 \sin \theta \cos \theta$$
$$= 2(-\frac{1}{4})(\sqrt{15}/4)$$
$$= -\sqrt{15}/8$$

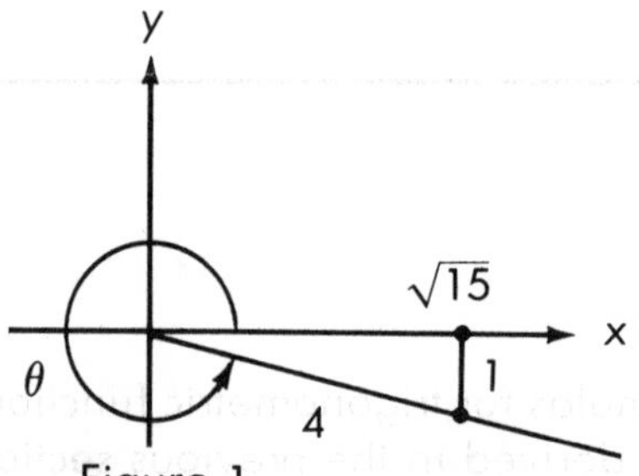

Figure 1

Now we will derive formulas for functions of $\frac{1}{2}\theta$. The following identities are called **half-angle formulas:**

$$\sin(\theta/2) = \pm\sqrt{(1-\cos \theta)/2} \tag{6}$$
$$\cos(\theta/2) = \pm\sqrt{(1+\cos \theta)/2} \tag{7}$$
$$\tan(\theta/2) = \pm\sqrt{(1-\cos \theta)/(1+\cos \theta)} \tag{8}$$

To prove Formula 6 we use Formula 3. We have

$$\sin^2\alpha = (1-\cos 2\alpha)/2$$

Writing $\alpha = \theta/2$ gives

$$\sin^2(\theta/2) = (1-\cos \theta)/2$$

and

$$\sin(\theta/2) = \pm\sqrt{(1-\cos \theta)/2}$$

Formulas 7 and 8 can be established in a similar manner and are left for the student.

EXAMPLE 4. Find sin 22.5° and tan 22.5° using the half-angle formulas.

Solution. Using Formula 6 and noting that sin 22.5° is positive, we have

$$\sin 22.5° = \sqrt{(1-\cos 45°)/2}$$
$$= \sqrt{(1-\sqrt{2}/2)/2}$$
$$= \tfrac{1}{2}\sqrt{(2-\sqrt{2})}$$

For tan 22.5°, we use Formula 8 with the positive root.

$$\tan 22.5° = \sqrt{(1-\cos 45°)/(1+\cos 45°)}$$
$$= \sqrt{(1-\sqrt{2}/2)/(1+\sqrt{2}/2)}$$
$$= \sqrt{(2-\sqrt{2})/(2+\sqrt{2})}$$

Self-test. Note that 112.5° = ½(225°) and find each of the following, using half-angle formulas.

(a) sin 112.5° = _____

(b) cos 112.5° = _____

(c) tan 112.5° = _____

EXAMPLE 5. If tan θ = −⅓ and θ is in quadrant II, find cos θ/2.

Solution. In Figure 2 we show θ in the standard position. We note that cos θ = −√8/3 and hence
$$\cos(θ/2) = \sqrt{(1+\cos θ)/2}$$
$$= \sqrt{[1+(-\sqrt{8}/3)]/2}$$
$$= \sqrt{(3-\sqrt{8})/6}$$

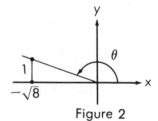

Figure 2

EXAMPLE 6. Prove that

$$\frac{\sin 2θ}{\sin θ} - \frac{\cos 2θ}{\cos θ} = \sec θ \qquad (9)$$

Solution. Starting with the more complicated side of Formula 9, we have

$$(\sin 2θ/\sin θ) - (\cos 2θ/\cos θ) = [(2 \sin θ\cos θ)/\sin θ] - [(2 \cos^2θ-1)/\cos θ]$$
$$= 2 \cos θ - 2 \cos θ + (1/\cos θ)$$
$$= 1/\cos θ = \sec θ$$

Recall that an identity is an equation that is true for all values for which both sides are defined. Thus Formula 9 is valid for all θ for which sin θ ≠ 0 and cos θ ≠ 0.

ANSWERS: (a) ½√2−√2   (b) −½√2+√2   (c) −√(2−√2)/(2+√2)

EXERCISES 3.2

In Exercises 1-14, use the half-angle formulas to compute the value of each expression.

1. $\cos(\pi/8)$ 2. $\csc(5\pi/8)$ 3. $\sin 15°$

4. $\cot 15°$ 5. $\sin 195°$ 6. $\cos(-195°)$

7. $\csc 22.5°$ 8. $\cos(\pi/12)$ 9. $\tan(\pi/12)$

10. $\sin(\pi/24)$ 11. $\tan(\pi/24)$ 12. $\cos(\pi/60)$

13. $\tan 7.5°$ 14. $\sin 52.5°$

15. Prove Identity 3.

16. Prove Identity 5.

17. Prove Identity 7.

18. Prove Identity 8.

In Exercises 19-26, simplify each expression.

19. $(\sin 2\theta)/(\sin \theta)$ 20. $(\sin 2\theta)/(2 \cos \theta)$

21. $2 \sin 3\theta \cos 3\theta$ 22. $2 \sin(\theta/2) \cos(\theta/2)$

23. $\cos^4\theta - \sin^4\theta$ 24. $(2 \tan 6\theta)/(1 - \tan^2 6\theta)$

25. $(\tan 3\theta - \tan 2\theta)/(1 + \tan 2\theta \tan 3\theta)$ 26. $\sqrt{(1 + \cos 4\theta)/2}$

27. If $\cos \theta = \frac{7}{9}$ and $3\pi/2 < \theta < 2\pi$, find $\sin(\theta/2)$ and $\cos(\theta/2)$.

28. If $\cos \theta = -\frac{7}{25}$ and $\pi < \theta < 3\pi/2$, find $\sin(\theta/2)$ and $\cos(\theta/2)$.

29. If $\sin \theta = -\frac{1}{8}$ and $\pi < \theta < 3\pi/2$, find $\sin(\theta/2)$ and $\cos(\theta/2)$.

30. If $\sin \theta = \frac{4}{5}$ and θ is in quadrant I, find $\sin 2\theta$, $\cos 2\theta$ and $\tan 2\theta$.

31. If $\cos \theta = -\frac{3}{5}$ and θ is in quadrant II, find $\sin 2\theta$, $\cos 2\theta$, and $\tan 2\theta$.

32. If $\sin \theta = -\frac{4}{5}$ and θ is in quadrant III, find $\sin 2\theta$, $\cos 2\theta$, and $\tan 2\theta$.

In Exercises 33-50, prove each identity.

33. $(\cos 2\theta + 1)\csc 2\theta = \cot \theta$ 34. $(2 \tan \theta)/(1 + \tan^2\theta) = \sin 2\theta$

35. $(\csc \theta)/(2 \cos \theta) = \csc 2\theta$ 36. $\sin 3\theta = 3 \sin \theta - 4 \sin^3\theta$

37. $\cos 3\theta = 4 \cos^3\theta - 3 \cos \theta$ 38. $(1 + \cos 2\theta)/(1 - \cos 2\theta) = \cot^2\theta$

39. $(1 - \cos 2\theta)/(\sin 2\theta) = \tan \theta$

40. $(\sin \theta - \cos \theta)^2 = 1 - \sin 2\theta$

41. $(\sin^2 2\theta)/(\sin^2\theta) = 4 - 4 \sin^2\theta$

42. $\sin 4\theta = 4 \cos \theta \sin \theta(1 - 2 \sin^2\theta)$

43. $\cot 2\theta = (\cot^2\theta - 1)/(2 \cot \theta)$

44. $2 \sin^2 2\theta + \cos 4\theta = 1$

45. $\cot \theta + \tan \theta = 2 \csc 2\theta$

46. $(\sin 3\theta/\sin \theta) - (\cos 3\theta/\cos \theta) = 2$

47. $(\sin \theta + \cos \theta)^2 = 1 + \sin 2\theta$

48. $[\cos \theta/(1 + \sin \theta)] - [(1 - \sin \theta)/\cos \theta] = 0$

49. $(\tan \theta + \sin \theta)/(2 \tan \theta) = \cos^2(\theta/2)$

50. $(2 - \sec^2\theta)/(2 \tan \theta) = \cot 2\theta$

3.3 THE INVERSE TRIGONOMETRIC FUNCTIONS

OBJECTIVES

1. Find all values of θ satisfying equations such as $\sin \theta = x$, where x is a number for which θ is one of the special angles.
2. Find the inverse relations of the trigonometric functions.
3. Define the inverse trigonometric functions and draw their graphs.
4. Evaluate expressions involving trigonometric functions.

In Section 1.5 we indicated that the inverse of a function f is a relation obtained by interchanging the first and second components of each ordered pair in f. We recall that if the function f is one-to-one, then the inverse function f^{-1} exists. In this section, we shall see that each of the trigonometric functions has an inverse relation, but these inverses are not functions. However, the inverse trigonometric functions exist on restricted intervals.

Let us consider the inverse of the sine function, $x = \sin \theta$. Suppose that θ is the radian measure of a variable angle. Then x is a variable number in the range $-1 \leq x \leq 1$. If

$$x = \sin \theta \qquad (1)$$

then we may express θ in terms of x as

$$\theta = \arcsin x \qquad (2)$$

read "θ is an arcsin of x" or "an inverse sine of x", or simply "θ is an angle whose sine is x". Many authors also use the notation

$$\theta = \sin^{-1} x \qquad (3)$$

In the notation $\sin^{-1}x$, -1 is not an exponent. Thus $\sin^{-1} \neq (\sin x)^{-1}$. The notations in Equations 2 and 3 indicate the same relation.

As an example, since $\sin(\pi/2)=1$, then $\pi/2$ is an angle whose sine is 1. We write

$$\arcsin 1 = \pi/2$$

Also, since $\sin(5\pi/2)=1$, then

$$\arcsin 1 = 5\pi/2$$

In fact, if $\theta \in S$, where

$$S=\{\theta \,|\, \theta = \pi/2 + 2n\pi, \, n \in Z\}$$

then

$$\arcsin 1 = \theta$$

Thus, we obtain more than one number for $\theta = \arcsin 1$. This relation is called a **multiple-value function**.

Self-test.

(a)   If $\theta = \arcsin(\tfrac{1}{2})$, then $\theta \in S =$ _____

(b)   If $\theta = \arcsin(-\sqrt{3}/2)$, then $\theta \in S =$ _____

Clearly, the sine function is not one-to-one and hence has no inverse function. However, we can restrict the domain over which the sine function is defined so that the restricted function is one-to-one. It is customary to choose the interval $[-\pi/2, \pi/2]$ and obtain a new function referred to as the Sine (read "cap sine") function. Thus we have

$$\text{Sin } \theta = \sin \theta, \quad -\pi/2 \leq \theta \leq \pi/2 \qquad (4)$$

The Sine and sine functions coincide on $[-\pi/2, \pi/2]$ and their graphs are identical on this interval (see Figure 1). The Sine function is one-to-one and has an inverse function.

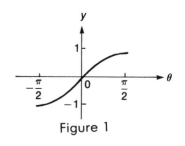

Figure 1

DEFINITION 3.1 The *inverse Sine function*, denoted by Arcsin, is defined by

$$\text{Arcsin } x = \theta \quad \text{if and only if} \quad x = \text{Sin } \theta \tag{5}$$

where $-1 \le x \le 1$ and $-\pi/2 \le \theta \le \pi/2$.

It is customary to refer to this function as the **Arcsine function**. The notation $\text{Sin}^{-1}x$ is frequently used in place of Arcsin x.

We recall that if f^{-1} is the inverse of the function f whose domain is X and range is Y, then

$$f^{-1}(f(x)) = x \qquad \text{for all } x \in X$$

and

$$f(f^{-1}(y)) = y \qquad \text{for all } y \in Y$$

Thus we obtain the following important identities.

$$\text{Arcsin}(\text{Sin } \theta) = \theta, \qquad -\pi/2 \le \theta \le \pi/2 \tag{6}$$

$$\text{Sin}(\text{Arcsin } x) = x, \qquad -1 \le x \le 1 \tag{7}$$

The graph of the Arcsine function is shown in Figure 2. In Figure 3 we show the graph of the relation $\theta = \arcsin x$.

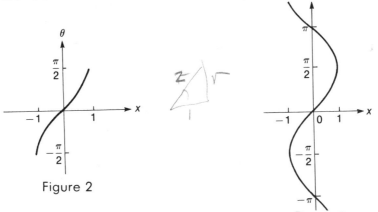

Figure 2

Figure 3

In defining the Sine function, we chose the interval $[-\pi/2, \pi/2]$. However, there are many alternative choices for an interval over which the sine function has an inverse function. Some examples are $[-3\pi/2, -\pi/2]$, $[\pi/2, 3\pi/2]$, and so on. In any such case, it is desirable to choose that interval for which the sine function maintains its full range of values.

EXAMPLE 1. Evaluate each of the following.

(a) $\text{Arcsin}(-\sqrt{2}/2)$ (b) $\text{Arcsin}(\frac{1}{2})$ (c) $\text{Arcsin}[\sin(\pi/4)]$ (d) $\text{Arcsin}[\cos(2\pi/3)]$

Solution.

(a) Let $\theta=\text{Arcsin}(-\sqrt{2}/2)$. Then, by definition,

$$\text{Sin } \theta = -\sqrt{2}/2$$

Now, since $\text{Sin}(-\pi/4)=-\sqrt{2}/2$ and $-\pi/2<-\pi/4<\pi/2$, we have $\theta=-\pi/4$. It should be noted that although $\sin(5\pi/4)=-\sqrt{2}/2$, $5\pi/4\neq\text{Arcsin}(-\sqrt{2}/2)$ since $5\pi/4\notin[-\pi/2, \pi/2]$.

(b) Let $\theta=\text{Arcsin}(\frac{1}{2})$. Then by definition,

$$\text{Sin } \theta = \frac{1}{2}$$

This implies $\theta=\pi/6$. That is

$$\text{Arcsin}(\frac{1}{2})=\pi/6.$$

(c) Using Identity 6, we have

$$\text{Arcsin}[\sin(\pi/4)]=\pi/4$$

(d) We have,

$$\text{Arcsin}[\cos(2\pi/3)]=\text{Arcsin}(-\frac{1}{2})=-\pi/6$$

Self-test. Evaluate each of the following.

(a) $\text{Arcsin}(\sqrt{2}/2)=$ _____

(b) $\text{Arcsin}(-\frac{1}{2})=$ _____

(c) $\text{Arcsin}(\sqrt{3}/2)=$ _____

(d) $\text{Arcsin}(\tan 3\pi/4)=$ _____

To define an inverse for the cosine function, we define the Cosine function (read "cap cosine") as follows.

$$\text{Cos } \theta=\cos \theta, \qquad 0\leq\theta\leq\pi \tag{8}$$

The Cosine function defined by Equation 8 is one-to-one and we have the following definition for its inverse.

DEFINITION 3.2 The *inverse Cosine function* (also called the *Arccosine function*), denoted by Arccos, is defined by

$$\text{Arccos } x=\theta \quad \text{if and only if} \quad \text{Cos } \theta=x \tag{9}$$

where $-1\leq x\leq 1$ and $0\leq\theta\leq\pi$.

The graph of $y=\text{Cos }\theta$ is shown in Figure 4. In Figure 5, we show the graph of $\theta=\text{Arccos }x$.

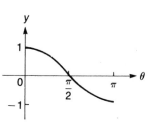

Figure 4

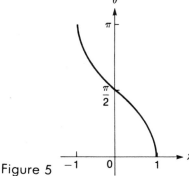

Figure 5

As in the case of the Arcsine functions, we have the following identities.

$$\text{Arccos(Cos }\theta)=\theta \qquad 0\le\theta\le\pi \tag{10}$$
$$\text{Cos(Arccos }x)=x, \qquad -1\le x\le 1 \tag{11}$$

EXAMPLE 2. Evaluate each of the following

    (a) $\text{Arccos}(\sqrt{2}/2)$ (b) $\text{Cos(Arccos }\tfrac{1}{2})$ (c) $\text{Cos(Arcsin }\tfrac{4}{5})$

Solution.

    (a)   Let $\theta=\text{Arccos}(\sqrt{2}/2)$. Then, by definition,

$$\text{Cos }\theta=\sqrt{2}/2, \qquad 0\le\theta\le\pi$$

Thus, $\theta=\pi/4$.

    (b)   From Identity 11, we have

$$\text{Cos(Arccos }\tfrac{1}{2})=\tfrac{1}{2}$$

    (c)   Let $\theta=\text{Arcsin}(\tfrac{4}{5})$. Then, by definition,

$$\text{Sin }\theta=\tfrac{4}{5}, \qquad -\pi/2\le\theta\le\pi/2$$

Since $\text{Sin }\theta$ is positive, then θ is the measure of an angle that lies in the first quadrant. Using the Pythagorean identity, we have:

$$\begin{aligned}
\text{Cos(Arcsin }\tfrac{4}{5}) &= \cos\theta \\
&= \sqrt{1-\text{Sin}^2\theta} \\
&= \sqrt{1-(\tfrac{4}{5})^2} \\
&= \sqrt{(9/25)} = \tfrac{3}{5}
\end{aligned}$$

Self-test. If $\theta=\text{Arccos}(\tfrac{1}{3})$, find

    (a) $\text{Cos }\theta=$ _____

    (b) $\text{Sin(arccos }\tfrac{1}{3})=$ _____

ANSWERS: (b) $\sqrt{1-(\tfrac{1}{3})^2}=2\sqrt{2}/3$ (a) $\tfrac{1}{3}$

For the remaining trigonometric functions, we have the following definitions:

$$\text{Tan } \theta = \tan \theta, \qquad -\pi/2 < \theta < \pi/2 \tag{12}$$

$$\text{Cot } \theta = \cot \theta, \qquad 0 < \theta < \pi \tag{13}$$

$$\text{Sec } \theta = \sec \theta, \qquad 0 \leq \theta \leq \pi, \ \theta \neq \pi/2 \tag{14}$$

$$\text{Csc } \theta = \csc \theta, \qquad -\pi/2 \leq \theta \leq \pi/2, \ \theta \neq 0 \tag{15}$$

For the functions defined in Equations 12-15, we define the inverse functions as follows:

DEFINITION 3.3

(a)   Arctan $x = \theta$ if and only if Tan $\theta = x$, where $x \in R$ and $-\pi/2 < \theta < \pi/2$. (See Figure 6).

(b)   Arccot $x = \theta$, if and only if Cot $\theta = x$, where $x \in R$ and $0 < \theta < \pi$. (See Figure 7).

(c)   Arcsec $x = \theta$, if and only if Sec $\theta = x$, where $|x| \geq 1$ and $0 \leq \theta \leq \pi$, $\theta \neq \pi/2$. (See Figure 8).

(d)   Arccsc $x = \theta$, if and only if Csc $\theta = x$, where $|x| \geq 1$ and $-\pi/2 \leq \theta \leq \pi/2$, $\theta \neq 0$. (See Figure 9).

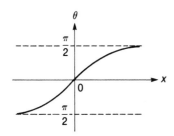

Figure 6

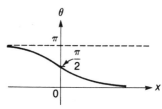

Figure 7

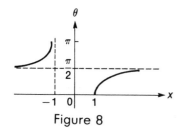

Figure 8

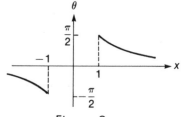

Figure 9

EXAMPLE 3. Evaluate each of the following.

(a) sec[Arctan($\frac{3}{5}$)]   (b) sin[Arccot($\frac{1}{3}$)$-$Arccsc($\frac{5}{3}$)]

Solution.

(a) Let $\theta=$Arctan($\frac{3}{5}$). Then by definition,

$$\text{Tan } \theta = \tfrac{3}{5}$$

Since Tan θ is positive, then θ is in the first quadrant. Therefore, (Recall that $1+\tan^2\theta=\sec^2\theta$)

$$\sec \theta = \sqrt{1+\text{Tan}^2\theta}$$
$$= \sqrt{1+(\tfrac{3}{5})^2}$$
$$= \sqrt{\tfrac{34}{25}} = \tfrac{1}{5}\sqrt{34}$$

(b) Let $\alpha=$Arccot($\frac{1}{3}$) and $\beta=$Arccsc($\frac{5}{3}$). Then

$$\text{Cot } \alpha = \tfrac{1}{3} \text{ and Csc } \beta = \tfrac{5}{3}$$

where α and β are the measures of the angles in the first quadrant. In Figures 10 and 11 we show α and β in the standard position.

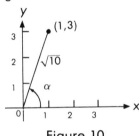

Figure 10

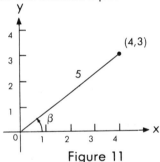

Figure 11

Thus we have

$$\sin[\text{Arccot}(\tfrac{1}{3})-\text{Arccsc}(\tfrac{5}{3})] = \sin(\alpha-\beta)$$
$$= \sin \alpha \cos \beta - \sin \beta \cos \alpha$$
$$= [(3/\sqrt{10})(\tfrac{4}{5})]-[(\tfrac{3}{5})(1/\sqrt{10})]$$
$$= 9/(5\sqrt{10})=(9\sqrt{10})/50$$

EXERCISES 3.3

In Exercises 1-12, find all values of each angle θ, where $0\le\theta\le2\pi$.

1. $\sin \theta=1$ 2. $\cos \theta=-1$ 3. $\cos \theta=\sqrt{3}/2$

4. $\sin \theta=-\frac{1}{2}$ 5. $\tan \theta=1$ 6. $\cot \theta=-1$

7. $\sin \theta=-\sqrt{2}/2$ 8. $\cos \theta=0$ 9. $\sec \theta=-2$

10. $\csc \theta=-1$ 11. $\cot \theta=-\sqrt{3}$ 12. $\sec \theta=-\sqrt{2}$

In Exercises 13-16, find all values of θ.

13. $\theta = \arcsin \frac{1}{2}$ 14. $\theta = \arccos (\sqrt{3}/2)$

15. $\theta = \arctan 0$ 16. $\theta = \text{arcsec}(-1)$

In Exercises 17-28, find the values of each angle.

17. $\text{Arccos}(\frac{1}{2})$ 18. $\text{Arcsin}(-\sqrt{3}/2)$ 19. $\text{Arcsin } 0$

20. $\text{Tan}^{-1}(-1)$ 21. $\text{Cot}^{-1}(\sqrt{3}/3)$ 22. $\text{Sec}^{-1}2$

23. $\text{Arccos}(-1)$ 24. $\text{Cos}^{-1}(-\sqrt{2}/2)$ 25. $\text{Sin}^{-1}(-\sqrt{2}/2)$

26. $\text{Arctan } \sqrt{3}$ 27. $\text{Arccot}(-1)$ 28. $\text{Arccos}(1/\sqrt{2})$

In Exercises 29-48, find the value of each expression and simplify.

29. $\text{Arccos}(\tan \pi)$ 30. $\text{Arcsin}(\sin\sqrt{7})$ 31. $\text{Arcsin}[\sin(3\pi/2)]$

32. $\text{Sin}^{-1}(\sin \pi/6)$ 33. $\text{Arccos}(\cos 3\pi/4)$ 34. $\text{Cos}^{-1}(\cos \pi/6)$

35. $\text{Arctan}(\tan 3\pi/4)$ 36. $\cos(\text{Arcsin } \frac{1}{2})$ 37. $\sin(\text{Tan}^{-1}1)$

38. $\cos[\text{Tan}^{-1}(-1)]$ 39. $\cos(\text{Sin}^{-1} \frac{3}{5})$ 40. $\sin(\text{Cot}^{-1} \frac{2}{3})$

41. $\cos(\text{Tan}^{-1}\sqrt{2}/3)$ 42. $\cot[\text{Cos}^{-1}(-\frac{1}{10})]$ 43. $\tan(\text{Sin}^{-1} \sqrt{3}/2)$

44. $\cos[\text{Arccot}(-\sqrt{3})]$ 45. $\sin[\text{Tan}^{-1}(-1)]$ 46. $\sec(\text{Arctan } \frac{2}{3})$

47. $\cos(\text{Arcsin } \frac{7}{25})$ 48. $\sin(\text{Arccos } \frac{5}{13})$

In Exercises 49-58, evaluate each expression.

49. $\sin[(\text{Tan}^{-1} \frac{1}{2})-(\text{Cos}^{-1} \frac{4}{5})]$ 50. $\sin[(\text{Sin}^{-1} \frac{1}{2})+(\text{Cos}^{-1} \frac{4}{5})]$

51. $\cos[(\text{Arcsin}\sqrt{3}/2)-(\text{Arccos } \frac{1}{2})]$ 52. $\cos[(\text{Sin}^{-1} \frac{1}{2})+(\text{Cos}^{-1} \frac{3}{5})]$

53. $\sin[(\text{Cos}^{-1}\sqrt{2}/2)+(\text{Sin}^{-1} \frac{4}{5})]$ 54. $\sin(2 \text{ Cos}^{-1} \frac{3}{5})$

55. $\sin(2 \text{ Arcsin } \frac{1}{2})$ 56. $\tan(\frac{1}{2} \text{ Sin}^{-1} \frac{3}{5})$

57. $\cos[2 \text{ Arcsin}(-\frac{4}{5})]$ 58. $\sin(\frac{1}{2} \text{ Tan}^{-1} \frac{15}{8})$

In Exercises 59-61, prove that each statement is true.

59. $\sin(\text{Cos}^{-1}x) = \sqrt{1-x^2}$ 60. $\text{Sin}^{-1}x = \text{Cos}^{-1} \sqrt{1-x^2}, \quad 0 \le x \le 1$

61. $\text{Arctan } \frac{2}{3} + \text{Arctan } \frac{1}{5} = \pi/4$

3.4 TRIGONOMETRIC EQUATIONS

OBJECTIVES

1. *Solve simple trigonometric equations not requiring the use of tables.*
2. *Use trigonometric tables.*
3. *Solve trigonometric equations requiring the use of tables.*

In earlier sections, we considered trigonometric identities, that is, equations that are true for all permissible values of the variables involved. Thus,

$$\sin^2\theta + \cos^2\theta = 1 \tag{1}$$

is a trigonometric identity while

$$2\cos\theta = 1 \tag{2}$$

is a trigonometric equation (called a **conditional equation**) that is not satified by all permissible values of the variable involved. For example, $\theta = \pi/2$ is a permissible value , but $\pi/2$ is not a solution of Equation 2. We find that $\theta = \pi/3$ is a solution of this equation. To solve trigonometric equations, we find all possible values of the variable that make the equation true.

EXAMPLE 1. Solve the equation $2\cos\theta = 1$.

Solution. We first solve for $\cos\theta$.

$$\cos\theta = \tfrac{1}{2}$$

Thus,

$$\theta = \arccos(\tfrac{1}{2})$$

or

$$\theta = \pi/3 + 2n\pi \text{ or } \theta = 5\pi/3 + 2n\pi$$

where *n* is any integer. Therefore, for Equation 2, we have the solution set

$$\{\theta \mid \theta = \pi/3 + 2n\pi, \text{ or } \theta = 5\pi/3 + 2n\pi, n \in Z\}$$

The following two examples illustrate the solution of trigonometric equations by factoring.

EXAMPLE 2. Solve the following equation:
$$\sin^2\theta = \sin\theta, \qquad 0 \le \theta < 2\pi \tag{3}$$

Solution. Equation 3 can be written as
$$\sin^2\theta - \sin\theta = 0$$
or
$$\sin\theta(\sin\theta - 1) = 0 \tag{4}$$

Equation 4 is satisfied if either factor is zero. Thus
$$\sin\theta = 0 \text{ or } \sin\theta - 1 = 0$$

The values of θ for which $0 \le \theta < 2\pi$ and $\sin\theta = 0$ are $\theta = 0$ and $\theta = \pi$. The value of θ for which $0 \le \theta < 2\pi$ and $\sin\theta = 1$ is $\theta = \pi/2$. Therefore, the solution set of Equation 3 is $\{0, \pi/2, \pi\}$.

EXAMPLE 3. Solve the following equation on $[0, 2\pi)$.
$$\cos 2\theta = \cos\theta \tag{5}$$

Solution. Using the double-angle formula we obtain
$$2\cos^2\theta - 1 = \cos\theta$$
or
$$2\cos^2\theta - \cos\theta - 1 = 0$$
Factoring yields
$$(2\cos\theta + 1)(\cos\theta - 1) = 0 \tag{6}$$
Thus, from Equation 6, we have
$$2\cos\theta + 1 = 0 \text{ or } \cos\theta - 1 = 0$$
That is,
$$\cos\theta = -\tfrac{1}{2} \text{ or } \cos\theta = 1$$

Therefore, the solution set of Equation 5 is given by $\{0, 120°, 240°\}$

Self-test. Solve the equation
$$\sin^2\theta = 1, \qquad 0 \le \theta < 360°$$
by filling in the blanks.

(a) $\sin\theta =$ _____ or $\sin\theta =$ _____.

(b) $\theta =$ _____ or $\theta =$ _____.

(c) The solution set is _____.

If a trigonometric equation involves more than one trigonometric function of the unknown, we express these functions in terms of a single function by use of identities. We illustrate this in the following example.

EXAMPLE 4. Solve the following equation on $[0, 2\pi)$.

$$\tan^2\theta + \sec\theta = 1 \tag{7}$$

Solution. We recall that $1 + \tan^2\theta = \sec^2\theta$. Thus, Equation 7 may be written as

$$\sec^2\theta - 1 + \sec\theta = 1$$

or

$$\sec^2\theta + \sec\theta - 2 = 0$$

or

$$(\sec\theta + 2)(\sec\theta - 1) = 0$$

Thus, $\sec\theta = -2$ or $\sec\theta = 1$. Therefore, the solution set of Equation 7 is $\{0, 2\pi/3, 4\pi/3\}$.

In solving trigonometric equations, it is best to check the resulting values of the unknown by substitution in the original equation. Certain operations on a given equation such as squaring or multiplying by a factor containing the unknown may introduce extraneous roots. Recall from algebra that an extraneous root is a value of the unknown that is obtained in the course of the work that does not satisfy the given equation.

EXAMPLE 5. Solve the following equation on $[0, 2\pi)$.

$$\tan\theta - \sec\theta = 0 \tag{8}$$

Solution. We write Equation 8 in terms of the sine and cosine to obtain

$$\frac{\sin\theta}{\cos\theta} - \frac{1}{\cos\theta} = 0$$

or

$$\frac{\sin\theta - 1}{\cos\theta} = 0 \tag{9}$$

Equation 9 is satisfied if $\sin\theta - 1 = 0$ or $\theta = \pi/2$. However, we find that $\theta = \pi/2$ does not satisfy Equation 8, since neither $\tan\theta$ nor $\sec\theta$ are defined. The solution set of Equation 8 is $\varnothing$ (the empty set).

EXAMPLE 6. From Table 1 in Appendix B, find each of the following.

      (a) Arccot 3.914   (b) Arctan 0.3727

Solution.

    (a)   We look in the body of the table in the column headed **cot** θ. We find 3.914 opposite $14°20'$. Thus we have

$$\text{Arccot } 3.914 = 14°20'$$

(b) In Table 1, we find tan 20°20′ = 0.3706 and tan 20°30′ = 0.3739. Thus, θ = Arctan 0.3727 lies between the angles 20°20′ and 20°30′. We set up the following arrangement.

$$10' \left[d \left[\begin{array}{l} \tan 20°20' = 0.3706 \\ \tan \theta\ \ = 0.3727 \end{array} \right] 0.0021 \atop \tan 20°30' = 0.3739 \right] 0.0033$$

We find

$$d/10 = (0.0021)/(0.0033) = {}^{7}\!/_{11}$$

or

$$d = {}^{7}\!/_{11}(10) \approx 6'$$

Thus,

$$\theta = \text{Arctan } 0.3727 = 20°26'$$

Self-test. Use Table 1 to fill in the blanks.

(a) Arctan 1.675 = _____, (b) tan 27°43′ = _____.

EXAMPLE 7. Solve the following equation on [0, 360°).

$$5 \cos \theta \cot \theta - 3 \cos \theta - 10 \cot \theta + 6 = 0 \qquad (10)$$

Solution. We factor the equation as follows.

$$\cos \theta (5 \cot \theta - 3) - 2(5 \cot \theta - 3) = 0$$

$$(5 \cot \theta - 3)(\cos \theta - 2) = 0$$

Thus, $5 \cot \theta - 3 = 0$ or $\cos \theta - 2 = 0$. The solution set of $\cos \theta = 2$ is $\varnothing$. Thus, the solution set of Equation 10 is the same as the solution set of

$$\cot \theta = {}^{3}\!/_{5} = 0.6000$$

We approximate the value of θ by interpolation.

$$10' \left[d \left[\begin{array}{l} \cot 59°00' = 0.6009 \\ \cot \theta\ \ = 0.6000 \end{array} \right] 0.0009 \atop \cot 59°10' = 0.5969 \right] 0.0040$$

We let d be the difference between θ and 59°00′. We have

$$(d/10) = \frac{0.0009}{0.0040}$$

or

$$d = {}^{9}\!/_{4} \approx 2'$$

ANSWERS:

Consequently, $\theta=59°2'$. Since $\cot \theta>0$, then θ could lie in quadrant I or III, and this gives us the second solution, $\theta=180+59°2'=239°2'$. Therefore, we see that the solution set of Equation 10 is $\{59°2',\ 239°2'\}$.

EXERCISES 3.4

In Exercises 1-4, find the solution set of each equation.

1. $2 \cos \theta+1=0$ 2. $\sin^2\theta-3=0$ 3. $\tan^2\theta=3$ 4. $\sec^2\theta-4=0$

In Exercises 5-28, find all solutions of each equation on $[0,\ 2\pi)$.

5. $2 \tan \theta \sin \theta+\tan \theta=0$

6. $4 \sin^2\theta=1$

7. $4 \cos^2\theta=3$

8. $\sin 2\theta-\sin \theta=0$

9. $\tan^2\theta=\tan \theta$

10. $2 \sin^2\theta+\sin \theta=0$

11. $\sec \theta \tan \theta+2 \tan \theta=0$

12. $\cos \theta+\cot \theta \cos \theta=0$

13. $2 \sin^2\theta-\sin \theta=0$

14. $2 \sin^2\theta-\cos \theta-1=0$

15. $(2 \cos \theta+1)(\tan \theta-1)=0$

16. $\tan \theta \sin \theta=\tan \theta$

17. $2 \sin \theta \cos \theta+2 \sin \theta-\cos \theta=1$

18. $\cot^2\theta+\csc \theta-1=0$

19. $\sin \theta \cos \theta=\frac{1}{4}$

20. $\cos 2\theta=\sin \theta$

21. $\cos 2\theta=\cos^2\theta-1$

22. $8 \cos^2\theta+2 \cos \theta=1$

23. $\tan^2\theta+2 \sec^2\theta-3=0$

24. $2 \csc^2\theta-\cot^2\theta=3$

25. $2 \tan \theta+\sqrt{3} \sin \theta \sec^2\theta=0$

26. $2 \sin \theta=\csc \theta+1$

27. $\sqrt{1-2 \tan^2\theta}+\tan \theta=0$

28. $3 \cot^2\theta+\sec^2\theta=5$

In Exercises 29-36, Use Table 1 in the Appendix to find the sine, cosine, and tangent of each angle.

29. $11°40'$ 30. $28°10'$ 31. $155°6'$ 32. $64°28'$

33. $27°14'$ 34. $35°53'$ 35. $32°13'$ 36. $77°7'$

In Exercises 37-40, find each angle correctly to the nearest minute.

37. Arctan 0.3057   38. Arccos 0.5299

39. Arcsin 0.6589   40. Arccos 0.5536

In Exercises 41-46, find all solutions of each equation on $[0, 360°)$.

41. $(\tan \theta - 2)(4 \sin \theta - 1) = 0$ 42. $6 \sin^2\theta - \cos \theta = 4$

43. $10 \sin^2\theta - 12 \sin \theta = 7$ 44. $\sin \theta + \cos \theta = 0$

45. $(5 \cos \theta - 2)(2 \sec \theta + 3) = 0$ 46. $9 \cos^2\theta + 3 \cos \theta - 2 = 0$

REVIEW OF CHAPTER THREE

1. What is the difference between a trigonometric identity and a conditional equation?

2. Write identities involving each of the following.

(a) $\sin(\theta_1 \pm \theta_2) =$ _____

(b) $\cos(\theta_1 \pm \theta_2) =$ _____

(c) $\sin 2\theta =$ _____

(d) $\cos 2\theta =$ _____

(e) $\tan(\theta_1 \pm \theta_2) =$ _____

(f) $\sin(\theta/2) =$ _____

(g) $\cos(\theta/2) =$ _____

3. Define the inverse trigonometric functions.

4. Without using tables, solve for θ.

    (a) $\theta = \arcsin(\sqrt{3}/2)$
    (b) $\theta = \text{Arccos}(-\frac{1}{2})$
    (c) $\theta = \arctan(-1)$
    (d) $\theta = \text{Arcsin}(-\sqrt{2}/2)$

5. Draw a graph of $\theta = \text{Arccos } x$.

6. Draw a graph of $\theta = \text{Arcsin } x$.

7. Draw a graph of $\theta = \text{Arctan } x$.

8. Evaluate each of the following:

    (a) $\cos(\text{Arcsin } \frac{4}{5})$
    (b) $\text{Cos}^{-1}[\cos(\pi/6)]$
    (c) $\sin[\text{Tan}^{-1}(\sqrt{2}/3)]$

9. Solve each of the following equations on $[0, 360°)$.

    (a) $2\sin^2\theta + \sin\theta = 0$
    (b) $2\cos 2\theta = 1$
    (c) $8\sin^2\theta + 2\sin\theta = 1$
    (d) $\tan^2\theta + \sec\theta = 1$
    (e) $\sin\theta + \cos\theta = 0$
    (f) $6\sin^2\theta - \cos\theta = 4$
    (g) $5\cos^2\theta - 2\cos\theta - 1 = 0$

TEST THREE

1. Use the Addition Formulas to write equivalent expressions.

    (a) $\tan(\alpha+\beta)=$ _____

    (b) $\cos(60°-45°)=$ _____

    (c) $\sin 67°\cos 43°-\cos 67°\sin 43°=$ _____

2. Find $\cos(\alpha+\beta)$ given that $\sin\alpha=\%$ and $\cos\beta=\%$ and α and β are in quadrant I.

3. Find $\sin(\pi/8)$.

4. Simplify the expression $(\sin 2\theta)/(\cos^2\theta)$.

5. Prove that $(\cot^2\theta-1)/(\cot\theta)=2\cot 2\theta$.

6. Evaluate $\operatorname{arccot}\sqrt{3}$.

7. Evaluate $\arcsin 0.3454$.

8. Evaluate $\operatorname{Arccos}\frac{1}{2}$.

9. Evaluate $\tan(\operatorname{Arctan}\%)$.

10. Simplify $\sin[\operatorname{Arccos}(-\frac{1}{2})+\operatorname{Arcsin}(\frac{1}{2})]$.

11. Solve the following equations on $[0, 2\pi)$.

    (a) $\sin 2\theta=\cos\theta$

    (b) $2\sin^2\theta+7\sin\theta-4=0$

    (c) $\sin\theta=\sin(2\theta-\pi)$

    (d) $10\sin^2\theta-12\sin\theta=7$

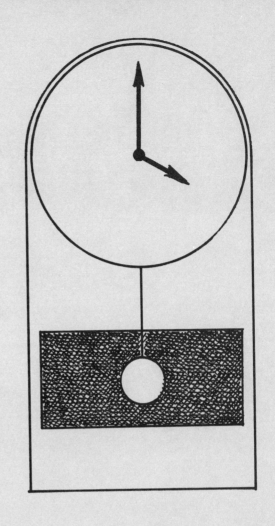

CHAPTER FOUR:
SOLVING TRIANGLES

4.1 SOLVING RIGHT TRIANGLES

OBJECTIVES

1 . *Solve right triangles.*
2 . *Solve applied problems involving right triangles.*

A triangle is described by three sides and three angles. When the measures of certain of these six fundamental elements are known, it is possible to compute the measures of the rest. The general problem of trigonometry is to **solve** triangles; that is, to find the length of each side and the measure of each angle of the triangle. Solving triangles is important in surveying, navigation, and in many other applications of trigonometry. The methods for solving triangles date back to the Babylonians. However, it was not until the sixteenth century that the trigonometric functions were defined (by Georg J. Rhaeticus) as ratios of the sides of a right triangle.

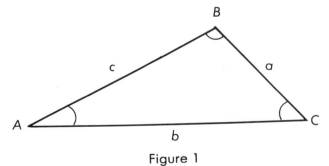

Figure 1

We shall begin by introducing the standard notation involving triangles. A triangle with angles A, B, and C is denoted by $\triangle ABC$. As indicated in Figure 1, the side opposite angle A is denoted by a, the side opposite the angle B is denoted by b, and the side opposite angle C is denoted by c. We note that the letters A, B, and

C play a triple role. Each represents (1) a vertex of the triangle (a point in the plane); (2) an angle; and (3) the measure of the angle. Similarly, each of a, b, and c represents a side of the triangle and the measure (or length) of the side. It is less confusing to use, for example, A with its triple role than to use three different symbols.

The definitions of the trigonometric functions of an acute angle A can be interpreted as ratios of a right triangle, one of whose angles is A, where $0 < A < \pi/2$. The triangle need not be in standard position on a coordinate system. In $\triangle ABC$ (a right triangle) shown in Figure 2, we have:

$$\sin A = \frac{\text{side opposite } A}{\text{hypotenuse}} = \frac{a}{c} \tag{1}$$

$$\cos A = \frac{\text{side adjacent } A}{\text{hypotenuse}} = \frac{b}{c} \tag{2}$$

$$\tan A = \frac{\text{side opposite } A}{\text{side adjacent } A} = \frac{a}{b} \tag{3}$$

Similar statements can be made about csc A, sec A, and cot A, which are the reciprocals of Equations 1, 2, and 3, respectively.

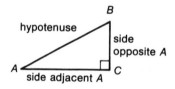

Figure 2   Figure 3

Since we already know one of the fundamental elements in a right triangle, namely the right angle, the triangle will be completely determined if we specifiy either two sides or an acute angle and one side.

EXAMPLE 1. Solve the right triangle ABC, given c=80 and A=20°, where C is the right angle.

Solution. First we find B (see Figure 3). We note that A+B+C=180°, so
$$B = 180° - (90° + 20°) = 70°$$
Now,
$$\sin A = \frac{a}{c}$$
or
$$\sin 20° = \frac{a}{80}$$
so that
$$a = 80 \sin 20° = 80(0.3420) = 27.36$$

Also,

$$\cos 20° = \frac{b}{80}$$

so that

$$b = 80 \cos 20° = 80(0.9397) = 75.18$$

Recall that an **angle of elevation** is the angle between the horizontal and the line of sight. If the line of sight is below the horizontal, the angle is called an **angle of depression** (see Figure 4).

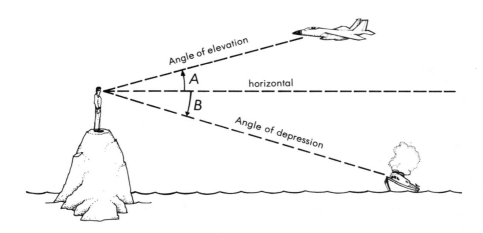

Figure 4

EXAMPLE 2. A beam of light is projected vertically making a spot on the clouds. A person 100 meters from the light source notes that the angle of elevation of the light spot on the clouds is 71°20′. What is the height of the cloud?

Solution. In Figure 5, we see that we have a right triangle in which h is to be determined. We have

$$\tan 71°20′ = h/100$$

Thus,

$$h = 100 \tan 71°20′ = 100(2.960) = 296$$

Hence, the cloud is approximately 296 meters above the ground. This method is used to measure cloud height at night.

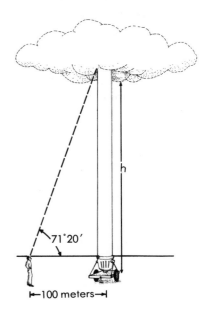

Figure 5

Self-test. An airplane is flying horizontally at an altitude of 2,000 meters. At a certain instant, the pilot sees a landing strip at an angle of depression of 22°20′. Assume that the landing strip and the airplane are in the same vertical plane and use Figure 6 to fill in the blanks.

(a) In Figure 6, the distance between the airplane and the landing strip is denoted by _____.

(b) $d=$_____ $\approx$_____ meters.

  The computation in Part b will be simplified if a calculator is available.

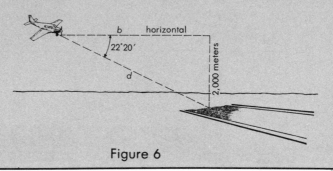

Figure 6

EXAMPLE 3. A statue on top of a hill is observed by a person at a point 500 meters distance horizontally from the vertical line of the statue. The observer notes that the angle of elevation of the top of the statue is 20°30′, and the angle of elevation of its base is 18°27′. How tall is the statue and how high is the base above the point of observation?

Solution. Let h_1 be the height of the statue and h_2 be the height of its base above the point of observation (see Figure 7).

$$\tan 18°27' = h_2/500 \tag{4}$$

and

$$\tan 20°30' = (h_1 + h_2)/500 \tag{5}$$

From Equation 4, we obtain $h_2 = 500 \tan 18°27' \approx 167.0$ and from Equation 5, we have $h_1 + h_2 = 500 \tan 20°30' \approx 187.0$. Thus, $h_1 \approx 187.0 - 167.0 = 20$. Therefore, the statue is approximately 20 meters tall and its base is about 167 meters above the point of observation.

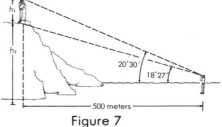

Figure 7

EXERCISES 4.1

In Exercises 1-15, solve each right triangle ABC from the data supplied (see the figure below).

1. $A=19°$, $c=70$

2. $B=43°$, $b=24$

3. $a=6$, $b=8$

4. $a=2$, $b=3$

5. $B=87°40'$, $b=9.7$

6. $A=67°40'$, $b=135$

7. $B=78°40'$, $a=134$

8. $A=82°20'$, $c=0.982$

9. $B=46°40'$, $c=0.0447$

10. $B=34.2°$, $c=135.4$

11. $A=25.6°$, $a=15$

12. $B=33.9°$, $b=56.02$

13. $B=23.7°$, $b=256.2$

14. $\sin B=\frac{4}{5}$, $c=20$

15. $\tan A=\frac{4}{3}$, $a=12$

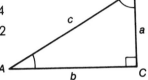

16. Find the area of each triangle in Exercises 1-4.

17. A balloon is held by a straight rope 150 meters long which is inclined at an angle 58° to the horizontal. Find the height of the balloon.

18. A guy wire to a steel tower makes an angle of 72°30' with the level ground. If the guy wire is perfectly straight and is anchored 5 meters from the base of the tower, how high up is the guy wire attached to the steel tower?

19. At a certain time of the day, a 15-meter mast casts a 20-meter shadow. What is the angle of elevation of the sun?

20. From an observation tower 20 meters high, the angle of depression to an object on level ground is 13°10'. How far from the foot of the tower is the object?

21. A guy wire is 12 meters long and is attached to a pole. The angle between the wire (which is perfectly straight) and the ground (which is horizontal) is 72°40'. How far from the base of the pole is the wire attached to the ground?

22. A 5-meter ladder leans against a wall, making a 70° angle with the ground. At what height does the upper end of the ladder touch the wall?

23. When the angle of elevation of the sun is 62°40', a tree casts a 15-meter shadow. What is the height of the tree?

24. An air traffic controller is in a tower that is 45 meters high. He locates a plane that is 1,300 meters away measured horizontally, at an angle of elevation of 15°50'. Find how high the plane is.

25. A forest ranger sees a fire from the top of a tower with an angle of depression of 5°34'. If the tower is 50 meters high, how far is the fire from the base of the tower?

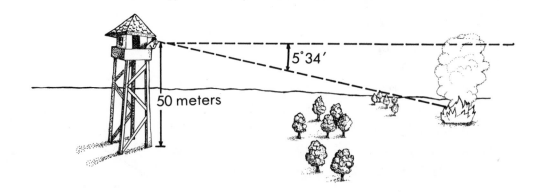

26. A blimp is 150 meters long. What angle does its length subtend at the eye of an observer 1,800 meters vertically below its center?

27. A boy scout out in the field wishes to estimate the width of a river. He climbs a tree 10 meters high above the ground and finds that the two banks of the river have angles of depression of 40° and 15°. How wide is the river?

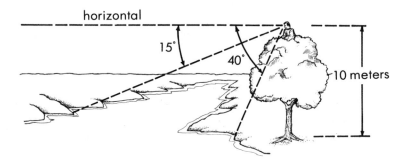

28. From an airplane flying 2 kilometers above the ocean, a pilot sees two ships directly to the east. The angles of depression to the ships are 8°20' and 75°40'. How far apart are the ships?

29. From the top of a building 60 meters high, the angles of depression of the top and base of a tree are 47°40' and 59°35' respectively. Find the height of the tree and its distance from the building.

30. When launched, a particular rocket rises vertically during the first few seconds of its flight. Three seconds after launch, an observer 1 kilometer away notes its angle of elevation is 6°40'. Five seconds later the angle is 62°10'. What is the distance traveled by the rocket during these five seconds?

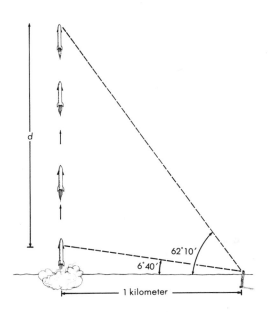

31. Two observers 1 kilometer apart spot an airplane. The angles of elevation of the airplane from the two observers are 22°50' and 24°40'. If the observers and the airplane are in the same vertical plane, compute the altitude of the airplane.

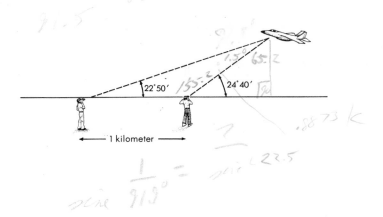

32. Two observation towers A and B are 15 kilometers apart, A being due south of B. The ranger in tower A observes a fire having bearing N40°30′E and the ranger at tower B observes the same fire at a bearing S30°20′E. Find the distance of the fire from each observation tower. Also find the shortest distance from the fire to the straight road joining towers A and B.

33. Ship A is due east of a lighthouse. Ship B is 15 kilometers due north of ship A. From ship B, the bearing to the lighthouse is S63°20′W. How far is ship A from the lighthouse? Ship B?

34. Prove that the area of right △ABC with right angle C is given by ¼c²sin 2A.

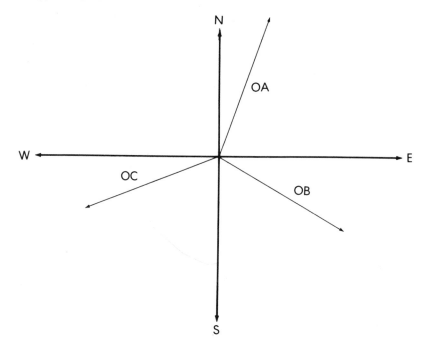

NOTE: In surveying and navigation, **bearings** (or directions) are given by reference to north or south by using an acute angle. Thus, in the figure above, the bearing of the line OA is N20°E, the bearing of OB is S60°E, and the bearing of OC is S70°W.

4.2 SOLUTION OF TRIANGLES:
THE SINE FORMULA

OBJECTIVES

1. *Prove the sine formula.*
2. *Use the sine formula to solve any triangle, given a side and two angles.*
3. *Use the sine formula to solve triangles given two sides and an angle opposite one of them.*
4. *Recognize the number of solutions when two sides and an angle opposite one of them are given.*

We now consider triangles that are not right triangles. Such triangles are called **oblique triangles**. In this and the next section, we state some useful results that permit us to solve any triangle when sufficient data is given. We note that when only one element of a triangle is known, say one side or one angle, a solution would exist. However, there are infinitely many triangles that contain a given element. If two elements are given, say two sides, one side and one angle, or two angles, and if a solution exists, it, too, would not be unique. In general, we must know at least three elements to insure uniqueness, although (as it will be shown below) even the knowledge of three elements is not sufficient to insure existence or uniqueness of a solution in every case.

In order to solve oblique triangles, we first derive the **sine formula** (or the **law of sines**). We shall consider any oblique triangle. It may or may not have an obtuse angle. Since the derivations are essentially the same, both cases will be considered simultaneously (see Figure 1). Let h be the length of a perpendicular from the vertex C to the opposite side c (extended if necessary as in Figure 1b). By the relations for right triangles, we have

$$\sin A = h/b \quad \text{or} \quad h = b \sin A \tag{1}$$

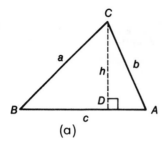

Figure 1

Also, from right $\triangle BDC$ in Figure 1a, we have

$$\sin B = h/a \quad \text{or} \quad h = a \sin B$$

and from right $\triangle BDC$ in Figure 1b, we have

$$h/a = \sin(\pi - B) = \sin B$$

So, in either case, we have

$$h = a \sin B \tag{2}$$

From Equations 1 and 2, it follows that

$$a \sin B = b \sin A$$

or

$$a/\sin A = b/\sin B \tag{3}$$

Similarly, by drawing the perpendicular from B to side b (extended if necessary), we can prove that

$$a/\sin A = c/\sin C \tag{4}$$

By combining the results in Equations 3 and 4, we obtain the sine formula. Thus, we have proved the following.

THEOREM 4.1   In any triangle ABC,

$$\frac{a}{\sin A} = \frac{b}{\sin B} = \frac{c}{\sin C} \tag{5}$$

We note that this formula applies for right triangles as well. However, it is usually easier to solve for a right triangle by the methods described in the previous section.

The sine formula is used to solve triangles when

1. two angles and one side are given, or
2. two sides and the angle opposite one of them are given.

EXAMPLE 1. Solve △ABC given that $a=5.4$, $A=34°$, and $C=67°$.

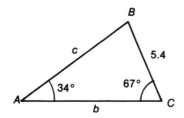

Figure 2

Solution. First we sketch △ABC (see Figure 2). Since $A+B+C=180°$, we find

$$B = 180° - (A+C)$$
$$= 180° - (34° + 67°) = 79°$$

Now we find the other two sides by using the sine formula. We have

$$b/\sin B = a/\sin A$$

Hence,

$$b = (5.4/\sin 34°)(\sin 79°) = (5.4/0.5592)(0.9816) \approx 9.5$$

Similarly,

$$c = (a/\sin A)(\sin C)$$
$$= (5.4/\sin 34°)(\sin 67°) = (5.4/0.5592)(0.9205) \approx 8.9$$

Thus, the unknown elements of the triangle have been found: $B=79°$, $b \approx 9.5$, and $c \approx 8.9$.

EXAMPLE 2. An observation tower 70 meters high is on a vertical cliff overlooking a bay. From the top of the tower, a ship is sighted with an angle of depression of 27°30′. The same ship is found to have an angle of depression 15°40′ measured from the base of the tower. Compute the distance of the ship from the cliff and find the height of the cliff.

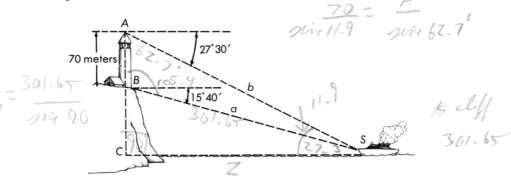

Figure 3

Solution. In Figure 3, let a be the length of the line segment from B to S, b being the length of the line segment from A to S, and $s=70$ meters be the length of the line segment from A to B (the height of the tower).

We are asked to find d(B,C) and d(C,S). In $\triangle ABS$, we have:

$$B = 15°40' + 90° = 105°40'$$
$$A = 90° - 27°30' = 62°30'$$
$$S = 180° - (105°40' + 62°30') = 11°50'$$

Using the sine formula, we find

$$a = (70/\sin 11°50')(\sin 62°30') = (70/0.2051)(0.8870) \approx 302.7$$

Next we consider right $\triangle BCS$ with right angle at C. We have

$$B = 90° - 15°40' = 74°20'$$

and

$$d(C,S) = a \sin B = (302.7)(\sin 74°20') \approx 291.4$$

Also,

$$d(B,C) = a \cos B = (302.7)(\cos 74°20') \approx 81.7$$

Thus, the cliff is about 81.7 meters high and the ship is 291.4 meters from the base of the cliff.

Self-test. Consider triangle ABC shown in Figure 4. Compute each of the following:

(a) C = _____

(b) $a = (b/\sin B)(\sin A)$ = _____

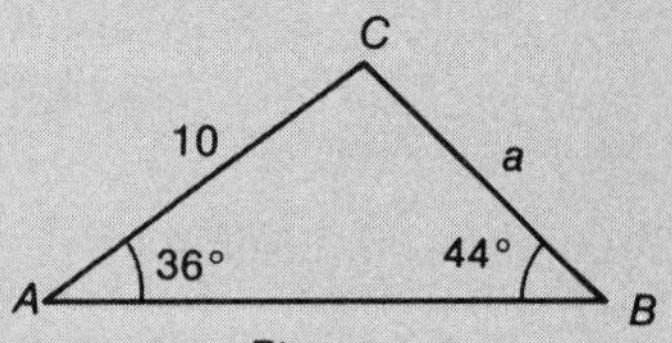

Figure 4

When two sides of a triangle and an angle opposite one of them are given, the sine formula may be used to solve the triangle. However, there may be

1.   no triangle possible with the given measures,
2.   only one triangle possible, or
3.   two triangles possible.

Because of these possibilities, this case is known as the **ambiguous case**.

Suppose in $\triangle ABC$, we are given a, b, and A, where A is an acute angle. The various possibilities are illustrated in the examples below.

EXAMPLE 3. Solve $\triangle ABC$ given that $a = 10$, $c = 20$, and $A = 30°$.

ANSWERS:

Solution. Using the sine formula (see Figure 5), we find

$$\sin C = (c \sin A)/a = (20 \sin 30°)/10 = 1$$

Since $0 < C < 180°$, we have $C = \text{Arcsin } 1 = 90°$. Hence

$$B = 180° - (90° + 30°) = 60°$$

and

$$b = (a/\sin A)(\sin B) = (10/\sin 30°)(\sin 60°) = 10\sqrt{3}$$

Therefore, there is a unique triangle in which $C = 90°$, $B = 60°$, and $b = 10\sqrt{3}$.

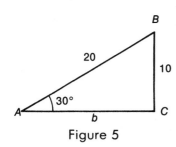

Figure 5

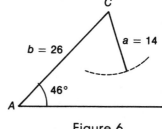

Figure 6

EXAMPLE 4. Solve $\triangle ABC$ given that $A = 46°$, $a = 14$, and $b = 26$.

Solution. First we look for B. Using the sine formula, we find

$$\sin B = (b \sin A)/a = (26 \sin 46°)/14 \approx 1.34$$

Since $-1 \le \sin B \le 1$, there is no solution. In Figure 6, we note that a is too short to form a triangle.

EXAMPLE 5. Solve $\triangle ABC$ given that $a = 15$, $b = 20$, and $A = 30°$.

Solution. First we solve for B. Using the sine formula, we find

$$\sin B = (b \sin A)/a = (20/15)(\sin 30°) \approx 0.667$$

Since $0° < B < 180°$, there are two values of B for which $\sin B = 0.667$. They are $41°50'$ and $138°10'$. Thus we have two possible solutions.

   Case 1. $B = 41°50'$. Then $C = 180° - (41°50' + 30°) = 108°10'$
and

$$c = (a/\sin A)(\sin C) = (15/\sin 30°)(\sin 108°10') \approx 28.5$$

These values form the triangle shown in Figure 7.

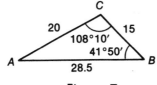

Figure 7

Case 2. $B = 138°10'$. Then $C = 180° - (30° + 138°10') = 10°50'$ and

$$c = (a/\sin A)(\sin C) = (15/\sin 30°)(\sin 11°50') \approx 6.2$$

These values form the second triangle, shown in Figure 8.

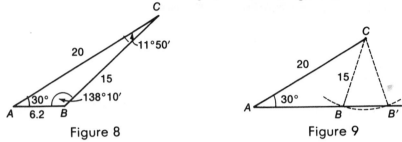

Figure 8   Figure 9

In Example 5, we found that an arc of radius $a = 15$ and center C meets the base at two points as shown in Figure 9. Hence, there are two solutions as indicated.

EXAMPLE 6. Solve $\triangle ABC$ given that $a = 20$, $b = 10$, and $A = 40°$.

Solution. We use the sine formula to solve for B.

$$\sin B = (b \sin A)/a = (10/20)\sin 40° = 0.3214$$

Since $0 < B < 180°$, we have

$$B = 18°45' \text{ or } B = 161°15'$$

We disregard $B = 161°15'$ because this triangle already has an angle of $40°$, and these two would total more than $180°$. (Recall in $\triangle ABC$, $A + B + C = 180°$.) Then

$$C = 180° - (40° + 18°45') = 121°15'$$

and

$$c = (a/\sin A)(\sin C) = (20/\sin 40°)(\sin 121°15') \approx 26.6$$

Therefore, there is a unique triangle in which $C = 121°15'$ and $c = 26.6$ (see Figure 10).

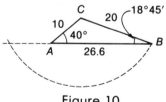

Figure 10

We now summarize our results. Consider $\triangle ABC$ and suppose that a, b, and A are given. Let h be the length of the line segment from C to the opposite side. Then $h = b \sin A$. We have the following cases.

Case 1. If $a = h$, then $\sin B = b[(\sin A)/a = h/a = 1$ and $B = 90°$. In this case, one solution exists and the resulting triangle is a right triangle (see Figure 11).

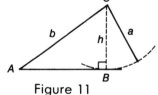

Figure 11

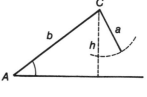

Figure 12

Case 2. If $a < h$, then $\sin B = [(\sin A)/a] < 1$. In this case, no triangle exists satisfying the given conditions (see Figure 12).

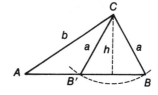

Figure 13

Case 3. If $a < b$ and $a > h$, then $\sin B = [(\sin A)/a] > 1$, and B has two values. In this case, there exist two triangles satisfying the given conditions (see Figure 13).

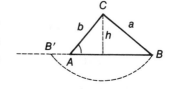

Figure 14

Case 4. If $a \geq b$, only one triangle (which is not a right triangle) exists satisfying the given conditions (see Figure 14). Note that the indicated angle A is not an angle of $\triangle AB'C$.

Self-test. Indicate the number of triangles that can be constructed with the following data. Identify the right triangles.

(a) $a=8$, $b=20$, $A=46°$ Ans. _____

(b) $a=3$, $b=4$, $A=30°$ Ans. _____

(c) $a=10$, $b=5$, $A=37°$ Ans. _____

(d) $a=3$, $b=3\sqrt{3}$, $A=30°$ Ans. _____

ANSWERS:
(a) None   (b) Two   (c) One   (d) One, right triangle

EXERCISES 4.2

In Exercises 1-16, solve $\triangle ABC$.

1. $A=30°$, $B=110°$, $b=12$ 2. $B=28°$, $C=120°$, $a=20$

3. $B=50°$, $A=20°$, $a=6$ 4. $A=32.6°$, $C=48.2°$, $b=5.7$

5. $A=106.5°$, $B=40.2°$, $a=7.5$

6. $A=36°$, $a=7$, $b=20$ 7. $A=30°$, $a=2$, $b=4$

8. $C=30°$, $b=17$, $c=6$ 9. $A=30°$, $a=6$, $b=20$

10. $A=43°$, $a=2$, $b=8$ 11. $B=30°$, $a=20$, $b=15$

12. $A=30°$, $a=3$, $b=5$ 13. $B=45°$, $a=4$, $b=3$

14. $A=43°50'$, $a=7.2$, $b=4.3$ 15. $C=51°40'$, $c=10.2$, $b=8.4$

16. $B=61°10'$, $a=30.3$, $b=24.2$

17. Show that the area of oblique triangle ABC is given by:

$$\text{Area of } \triangle ABC = \tfrac{1}{2}bc \sin A$$

$$= \tfrac{1}{2}ac \sin B$$

$$= \tfrac{1}{2}ab \sin C$$

18. Find the area of each triangle in Exercises 1-4 above. (See Exercise 17.)

19. A vertical building stands on a street in San Francisco that slopes downward at an angle of $7°40'$. From a point 50 meters down this street from the base of the building, the angle of elevation of the building is $58°20'$. Compute the height h of the building.

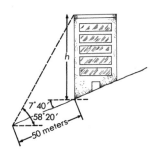

20. A yacht starts from a point A and sails 8 miles due south and then 6 miles on a course S20°E. How far is the yacht from the starting point A?

21. Two straight roads diverge from a point A at an angle of 32°. Juanita and Carlos leave A at the same time. Carlos walks at 5 kilometers per hour on one road and Juanita bicycles at 14 kilometers per hour along the other. How far apart are Juanita and Carlos at the end of 1 hour?

22. The angle of elevation of the top of a building is 30° from one point A and 47° from another point B, 25 meters nearer the base of the building which is in line with A and B and at the same level. How high is the building? (Hint: First find d(B,P) or d(A,P)

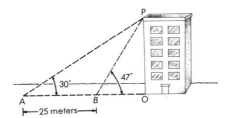

23. Two markers A and B are on opposite sides of a crater on Mars. A third marker C is 100 meters from A. If A=110° and C=38°, compute the width of the crater.

24. Two air controllers A and B, 10 kilometers apart, spot an airplane between them. The angles of elevation from A and B are 50°20′ and 33°40′, respectively. Assuming that A,B, and the airplane are in the same vertical plane, how far is the airplane from the nearest controller?

25. A ship in a fog is S20°E of Station A and S50°W of Station B. If Station A is 50 kilometers due west of Station B, how far is the ship from B?

26. The diagonals of a parallelogram are 20 centimeters and 30 centimeters long and they intersect at an angle of 25°. Find the sides of the parallelogram.

27. A tree is 22 meters tall and is growing vertically on a sloping hill. From a point 52 meters from the base of the tree (measured straight downhill), the tree subtends an angle of 20°40′. Find the angle that the sloping hill makes with the horizontal plane.

28. A man 6 feet tall stands in a vertical position on a hillside of slope 25°. What is the length of his shadow when the sun's elevation is 56°?

4.3 SOLUTION OF TRIANGLES:
THE COSINE FORMULA

OBJECTIVES

1. *Derive the cosine formula.*
2. *Use the cosine formula to solve triangles given three sides.*
3. *Use the cosine formula to solve triangles given two sides and the included angle.*
4. *Derive and use Heron's formula for the area of a triangle.*

There are cases where a triangle that cannot be solved using the sine formula can be solved with an additional law known as the **cosine formula** (or the **law of cosines**). For example, if the three sides of a general triangle are known, the sine formula cannot be used to solve the triangle.

We now derive the cosine formula. Consider $\triangle ABC$. We shall place one of the vertices, say A, at the origin of a rectangular coordinate system and place vertex B on the positive x-axis. Then the coordinates of B are $(c,0)$ and the coordinates of C are $(b \cos A, b \sin A)$. It is immaterial whether A is acute as in Figure 1a, or obtuse, as in Figure 1b. Using the distance formula, we find:

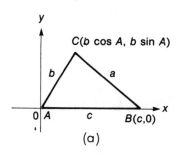
(a)

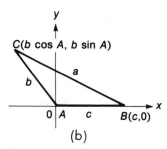
(b)

Figure 1

$$[d(B,C)]^2 = (b \cos A - c)^2 + (b \sin A - 0)^2$$
$$= b^2\cos^2 A - 2bc \cos A + c^2 + b^2\sin^2 A$$
$$= b^2(\cos^2 A + \sin^2 A) - 2bc \cos A + c^2$$
$$= b^2 + c^2 - 2bc \cos A$$

But $[d(B,C)]^2 = a^2$, and so

$$a^2 = b^2 + c^2 - 2bc \cos A$$

Had we placed one of the other vertices at the origin, we would have obtained

$$b^2 = a^2 + c^2 - 2ac \cos B$$

or

$$c^2 = a^2 + b^2 - 2ab \cos C$$

We have proved the following result.

THEOREM 4.2   In $\triangle ABC$

$$a^2 = b^2 + c^2 - 2bc \cos A \qquad\qquad (1)$$
$$b^2 = a^2 + c^2 - 2ac \cos B \qquad\qquad (2)$$
$$c^2 = a^2 + b^2 - 2ab \cos C \qquad\qquad (3)$$

The student should note that the cosine formula is a generalization of the Pythagorean theorem.

EXAMPLE 1. Solve $\triangle ABC$ given that $a=7.1$, $b=5.4$, and $c=3.5$.

Solution. Using Equation 1, we find: ·

$$\cos A = (b^2 + c^2 - a^2)/(2bc)$$
$$= [(5.4)^2 + (3.5)^2 - (7.1)^2]/[2(5.4)(3.5)]$$
$$= -9.0/37.8 \approx -0.2381$$

Since $\cos A < 0$, A is an obtuse angle. Using Table 1, we find $A = 103°47'$. Similarly, we find

$$\cos B = (a^2 + c^2 - b^2)/(2ac)$$
$$= [(7.1)^2 + (3.5(^2 - (5.4)^2]/[2(7.1)(3.5)] \approx 0.6740$$

From this, $B = 47°37'$. Lastly, $C = 180° - (103° + 47°37') = 28°36'$.

We note that when three sides of a triangle are given, there is either a unique solution or no solution. For instance, if the sum of any two sides is equal to or less than the third side, no triangle exists.

When two sides and the angle between them are known, a unique solution always exists and one can solve such a triangle using the cosine formula.

EXAMPLE 2. Solve $\triangle ABC$ given that $a=3$, $c=4$, and $B=110°$.

Solution. Using the cosine formula, we find:

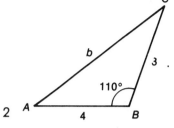

$$b^2 = a^2 + c^2 - 2ac \cos B$$
$$= 3^2 + 4^2 - 2(3)(4)\cos 110°$$
$$= 33.208$$

Then,

$$b = \sqrt{33.208} \approx 5.76$$

Figure 2

Next, we need to find angles A and C (see Figure 2). Here we can use either the sine or the cosine formulas. Using the former, we find

$$\sin A = (a \sin B)/b$$
$$= (3 \sin 110°)/5.76 \approx 0.4894$$

Thus,

$$A = 29°18'$$

Lastly, $C = 180° - (110° + 29°18') = 40°42'$.

In Table 1, we indicate the specific method of solution for given data.

DATA GIVEN	ABBREVIATION	METHOD OF SOLUTION
1. Two angles and one side	ASA	sine formula
2. Two angles and an angle opposite one of them	SSA	sine formula
3. Two side and included angle	SAS	cosine formula
4. Three side	SSS	cosine formula
5. Three angles	AAA	cannot be solved

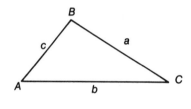

Figure 3

Table 1

We now derive a formula for finding the area of $\triangle ABC$ (see Figure 3) when three sides are given. First, we recall (see Exercise 17 in Section 4.2) that the area A of $\triangle ABC$ is given by

$$A = \tfrac{1}{2}bc \sin A \qquad (4)$$

Squaring, we get

$$A^2 = \tfrac{1}{4}b^2c^2\sin^2 A \qquad (5)$$

$$= \tfrac{1}{4}b^2c^2(1 - \cos^2 A) \qquad (6)$$

From Equation 1, we have

$$\cos A = (b^2 + c^2 - a^2)/2bc$$

Thus

$$\cos^2 A = (b^2 + c^2 - a^2)^2/4b^2c^2$$

and

$$1 - \cos^2 A = 1 - [(b^2 + c^2 - a^2)^2]/4b^2c^2$$

$$= [4b^2c^2 - (b^2 + c^2 - a^2)^2]/4b^2c^2$$

Letting $1-\cos^2 A = \sin^2 A$ and factoring the right-hand side, we get

$$\sin^2 A = [2bc-(b^2+c^2-a^2)][2bc+(b^2+c^2-a^{2}]/4b^2c^2$$
$$= [a^2-(b-c)^2][(b+c)^2-a^2]/4b^2c^2$$

Factoring again, we find:

$$\sin^2 A = [(a-b+c)(a+b-c)(b+c-a)(b+c+a)]/4b^2c^2 \qquad (7)$$

Now let $s = \frac{1}{2}(a+b+c)$, which is half the perimeter of $\triangle ABC$. Then

$$a+b+c=2s$$
$$a+b-c=2(s-c)$$
$$b+c-a=2(s-a)$$
$$a+c-b=2(s-b)$$

Substituting these in Equation 7 gives:

$$\sin^2 A = [2s \times 2(s-a) \times 2(s-b) \times 2(s-c)]/4b^2c^2$$
$$= [4s(s-a)(s-b)(s-c)]/b^2c^2$$

and

$$\sin A = (2/bc)\sqrt{s(s-a)(s-b)(s-c)} \qquad (8)$$

Substituting Equation 8 into Equation 4, we get

$$A = \sqrt{s(s-a)(s-b)(s-c)} \qquad (9)$$

This formula is called **Heron's formula** (or **Hero's Formula**) for the area of a triangle.

EXAMPLE 3. Find the area of $\triangle ABC$, given that $a=4$, $b=7$, and $c=9$.

Solution. First we find half the perimeter.

$$s = \frac{1}{2}(a+b+c) = \frac{1}{2}(4+7+9) = 10$$

Using Heron's formula, we find the area A, of $\triangle ABC$.

$$A = \sqrt{s(s-a)(s-b)(s-c)}$$
$$= \sqrt{10(10-4)(10-7)(10-9)}$$
$$= \sqrt{180} \approx 13.42$$

HERON OF ALEXANDRIA

Heron of Alexandria lived around 100 B.C. He was an accomplished surveyor, engineer, and mathematician. Although the formula for the area of a triangle, is credited to him, it was Archimedes who is believed to have proved it first. Heron applied mathematics to the design of such items as theatres, baths, measuring instruments, and war engines. Heron's writings on geodesy were used for hundreds of years.

EXAMPLE 4. Find the three altitudes of the triangle in Example 3.

Solution. (See Figure 4.) First we note that the area of a triangle is given by

$$\text{Area} = \tfrac{1}{2}(\text{base})(\text{altitude})$$

Hence, we have

$$A = \tfrac{1}{2}ch$$

or

$$h_1 = 2A/c = 2(13.42)/9 \approx 2.98$$

Similarly,

$$h_2 = 2A/a = 2(13.42)/4 \approx 6.71$$
$$h_3 = 2A/b = 2(13.42)/7 \approx 3.83$$

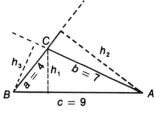

Figure 4

EXERCISES 4.3

In Exercises 1-12, solve $\triangle ABC$ (if a solution exists). Compute the angle measures to the nearest minute.

1. $a=4,\ b=7,\ c=8$

2. $a=6.1,\ b=5.4,\ c=4.2$

3. $a=6,\ b=8,\ c=4$

4. $a=10,\ b=8,\ c=16$

5. $a=6.2,\ b=3.5,\ c=4.2$

6. $a=2.3,\ b=4.2,\ c=2.5$

7. $a=12,\ c=24,\ B=30°$

8. $A=60°,\ b=5,\ c=4$

9. $A=110°,\ b=21,\ c=15$

10. $a=5,\ b=8,\ C=60°$

11. $B=135°,\ a=5\sqrt{2},\ c=7$

12. $C=23°30',\ b=52,\ c=18$

In Exercises 13-17, use Heron's formula to find the area of $\triangle ABC$.

13. $a=3,\ b=4,\ c=6$

14. $a=15,\ b=17,\ c=10$

15. $a=7,\ b=9,\ c=12$

16. $a=21.2,\ b=32.6,\ c=40.8$

17. $a=8.1,\ b=8.9,\ c=10.6$

18. Find the altitudes of the triangles in Exercises 13-16.

19. Find the area of a parallelogram whose sides have lengths 9 and 12 and which has one angle of 35°.

20. Find the area of the rhombus whose sides have length 10 with one angle of 42°.

21. Two sides of a triangle are 25 centimeters and 46 centimeters and the area is 490 square centimeters. Compute the measure of the included angle to the nearest minute. Is the solution unique?

22. A surveyor wishes to calculate the distance between two trees B and C across a river from where he is standing. If the distance from his position, A, to B is 70 meters, the distance from A to C is 100 meters, and the angle of A (see Figure below) is 56°, compute the distance between the two trees.

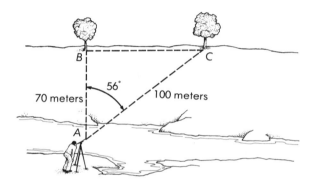

19. Find the area of a parallelogram whose sides have lengths 9 and 12 and which has one angle of 35°.

20. Find the area of the rhombus whose sides have length 10 with one angle of 42°.

21. Two sides of a triangle are 25 centimeters and 46 centimeters and the area is 490 square centimeters. Compute the measure of the included angle to the nearest minute. Is the solution unique?

22. A surveyor wishes to calculate the distance between two trees B and C across a river from where he is standing. If the distance from his position, A, to B is 70 meters, the distance from A to C is 100 meters, and the angle of A (see Figure below) is 56°, compute the distance between the two trees.

23. An airplane leaves an airport and flies 200 kilometers east. It then flies 300 kilometers in the direction N22°W. How far is the plane from its starting point?

24. Two ships leave port simultaneously. Ship A sails N25°E at 20 knots and ship B sails N28°W at 25 knots. How far apart are the ships 2 hours later? 3 hours later?

25. The Leaning Tower of Pisa was 54.60 meters high when first constructed. It now leans at an angle of 84.5° from the horizontal. At a certain time of the day when the sun was behind the tower (away from the tilt), the length of its shadow was 42 meters. Calculate the angle of elevation of the sun at that time.

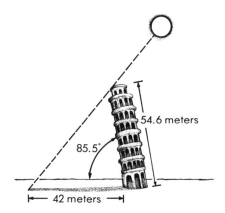

54.6 meters

85.5°

├── 42 meters ──┤

26. A soccer goal is 24 feet wide. A player shoots the ball when he is 54 feet from one goal post and 60 feet from the other. Calculate the angle within which the player must execute a ground shot to score.

27. The parallel sides of a trapezoid are 4 and 10 centimeters long and the obliques sides are 8 and 12 centimeters long. Find the angles and the area of the trapezoid.

28. A wire fence 550 meters long is set up to enclose a triangular lot. One side is 150 meters long and another is 200 meters long. Find the area and the angles of the triangle enclosed.

29. Prove that in $\triangle ABC$,
$$\frac{a-b}{a+b} = \frac{\tan \frac{1}{2}(A-B)}{\tan \frac{1}{2}(A+B)}$$

This is called the **tangent formula** (or **law of tangents**).

30. Use the tangent formula to solve Exercises 7-12.

REVIEW OF CHAPTER FOUR

1. The six elements of $\triangle ABC$ are _____.

2. What is meant by solving $\triangle ABC$?

3. Using right $\triangle ABC$, define the six trigonometric functions for $A(0<A<\pi/2)$.

4. The sum of the measures of the angles of a triangle is _____.

5. The angle between the horizontal and the top of a mountain is called _____.

6. The angle between the horizontal and the bottom of the valley is called _____.

7. In right $\triangle ABC$, C is a right angle. Solve the triangle given that $a=4.6$ and $c=7.5$.

8. In right $\triangle ABC$ one side bears east. If the hypotenuse is 250 meters long and it bears N46°20′E, find the perimeter.

9. A triangle that is not a right triangle is called _____.

10. State the sine formula.

11. Explain the ambiguous case that may arise in solving triangles.

12. In $\triangle ABC$, $A=30°$, $a=35$, and $b=50$. Solve the triangle.

13. The base angles of an isosceles triangle measure 46°40′ and the base is 50 meters long. Find the perimeter.

14. Consider $\triangle ABC$ and suppose a, b, and A are given. Discuss the different possible solutions.

15. In $\triangle ABC$, if A, b, and c are given, then the area of the triangle is _____.

16. State and prove the cosine formula.

17. Solve $\triangle ABC$ given that $a=6$, $c=8$, and $B=112°10′$.

18. Indicate the method of solution with the given data.

        a. ASA: _____

        b. SSA: _____

        c. SAS: _____

        d. SSS: _____

19. If three sides of $\triangle ABC$ are given, then the area of the triangle is given by _____.

20. Find the area of an equilateral triangle if the measure of each side is 20 cm.

TEST FOUR

1. In right triangle ABC, if $C=90°$, $B=15°$, and $b=8$ cm, find c.

2. In $\triangle ABC$, $a=20$, $b=15$, and $A=30°$. Use the sine formula to solve the triangle.

3. In $\triangle ABC$, $a=5$, $b=10$, $c=8$. Solve the triangle.

4. Find the area of the triangle in Problem 3.

5. A person is standing at point A on a beach and sights a boat and a lighthouse at points B and C respectively. If angle C is $90°$, angle A is $60°$ and the distance from A to C is 100 meters, how far is the boat from the lighthouse? From the person?

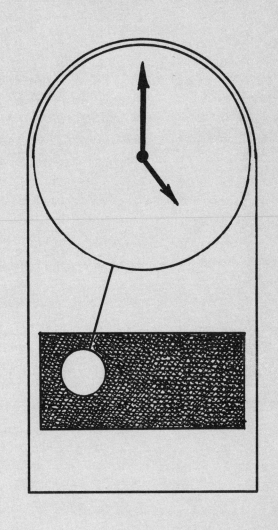

CHAPTER FIVE:
COMPLEX NUMBERS

5.1 THE COMPLEX NUMBERS

OBJECTIVES

1. Define the set of complex numbers.
2. Define equality, addition, subtraction, multiplication and division for complex numbers.
3. Establish that the set of real numbers is a subset of the set of complex numbers.

Many mathematical and scientific problems are adequately solved using real numbers. However, there are many problems that lead to equations that have no solutions in the real number system R. Indeed, the equation $x^2 + 1 = 0$ has no solution in R. Fortunately, it is possible to construct a number system, called the **system of complex numbers**, which contains the real numbers and which has the property that equations such as $x^2 + 1 = 0$ do have solutions.

In dealing with the real numbers we do not define $\sqrt{x}$ for $x < 0$. For instance, what does the expression $\sqrt{-1}$ stand for?. The symbol $\sqrt{-1}$ stands for some number which, when multiplied by itself, gives -1. Euler used the symbol

$$i = \sqrt{-1}$$

Then $i^2 = -1$. What kind of number is i? If it were a real number, then it would be negative, positive or zero.

NOTE: The terms **real number** and **imaginary number** should not mislead the reader. Imaginary numbers exist just as much as real numbers. Before the rational numbers were invented, 5/3 was called a "nonreal" number; before the irrationals were invented, $\sqrt{2}$ was considered a "nonreal" number; and as for negative numbers, although the Arabs introduced them to Europe, mathematicians of the sixteenth and seventeenth centuries did not accept them as real numbers.

If i were zero, then $i^2 = 0$. This is certainly not -1, and so $i \neq 0$.

If i were negative, then i^2 would be positive, since a negative number times a negative number is a positive number. But $i^2 = -1$, which is negative, so i cannot be a negative number.

Finally, if i were positive, then i^2 would be positive, since a positive number times a positive number is a positive number. But $i^2 = -1$, which is negative. Therefore, i is not a positive number. Thus, i is not a real number.

Descartes used the term **imaginary number** and, in 1832, Gauss* used the term **complex number** to refer to numbers such as i. Such numbers play a very important role in mathematics and the sciences. Complex numbers are indispensable in such areas as electrical engineering, aeronautical engineering, elasticity, and heat conduction.

We are all probably familiar with complex numbers written in the form $a + bi$ where a and b are real numbers and $i = \sqrt{-1}$. We shall use this form to define the complex number system. Then we shall use the Cartesian product to define a complex number as an ordered pair (a,b) of real numbers and relate this to the representation $a + bi$. The ordered pair representation will make it possible to establish the connection between a complex number $a + bi$ and the point (a,b) in the complex plane (see Section 5.2).

The set C of complex numbers, an extension of the real number system, consists of all expressions of the form $a + bi$ with the definitions of equality, sum and product given below. It should be noted that by defining a complex number in the form $a + bi$, the $+$ sign is not interpreted as the symbol for addition but only as a part of the notation for a complex number.

In the complex number $z = a + bi$, the number a is called the **real part** of z, and we write $a = Re(z)$; the number b is called the **imaginary part** of z, and we write $b = Im(z)$. If $b \neq 0$, then z is called an **imaginary number**.

CARL FRIEDRICH GAUSS

Carl Friedrich Gauss (1777-1855), born in Brunswick, Germany, was one of the greatest mathematicians of all time. His outstanding talent manifested itself at an early age, but unfortunately, Gauss' father, a gardener and bricklayer had neither the inclination nor means to educate his son. Gauss' abilities came to the attention of the Duke of Brunswick who supported Friedrich through school and sent him to the University of Gottingen.

Gauss was a prolific writer who tackled many areas in mathematics. His early reputation was made in the theory of numbers. He invented the method of least squares, contributed to the theory of probability, and rediscovered the asteroid Ceres after a number of other prominent astronomers had failed to do so. He was appointed professor of astronomy and director of the observatory in Gottingen. Although Gauss disliked the administrative duties and considered teaching a waste of time, he was an outstanding teacher. Among his more famous students were R. Dedekind and G.F. Riemann.

DEFINITION 5.1 If $z=a+bi$ and $w=c+di$ are two complex numbers, then z is equal to w (we write $z=w$) if and only if

$$a=c \quad \text{and} \quad b=d.$$

Thus, the complex number $2+5i$ and $2+\sqrt{25}i$ are equal, whereas $1+4i$ and $3+4i$ are not equal.

Now, if ordinary rules of arithmetic are to be valid, then complex numbers must add as follows:

$$\begin{aligned}
(a+bi)+(c+di) &= a+c+bi+di \\
&= (a+c)+(b+d)i
\end{aligned}$$

We take this as the definition of addition in the set C.

Similarly, if we agree on $i^2=-1$, then we want multiplication to operate as follows:

$$\begin{aligned}
(a+bi)(c+di) &= ac+adi+cbi+bdi^2 \\
&= (ac+bdi^2)+(ad+cb)i \\
&= (ac-bd)+(ad+cb)i
\end{aligned}$$

Formally, we have

DEFINITION 5.2 Let $z=a+bi$ and $w=c+di$ be two complex numbers. Then their *sum* and *product* are defined by

(i) $z+w=(a+bi)+(c+di)=(a+c)+(b+d)i$

(ii) $zw =(a+bi)(c+di)=(ac-bd)+(ad+cb)i$

EXAMPLE 1. If $z_1=-1+3i$ and $z_2=2-5i$, compute each of the following.

(a) z_1+z_2 (b) z_1z_2

Solution.

$$\begin{aligned}
\text{(a) } z_1+z_2 &= (-1+3i)+(2-5i) \\
&= (-1+2)+(3-5)i \\
&= 1-2i
\end{aligned}$$

$$\begin{aligned}
\text{(b) } z_1z_2 &= (-1+3i)(2-5i) \\
&= (-1)(2)+(-1)(-5i)+(3i)(2)+(3i)(-5i) \\
&= -2+5i+6i-15i^2 \\
&= -2+15+11i \\
&= 13+11i
\end{aligned}$$

Self-test. Let $z_1 = 2 - 4i$ and $z_2 = -5 + 7i$. Then

(a) $z_1 + z_2$ = _____

(b) $z_1 z_2$ = _____

From Definition 5.2 and properties of real numbers, it follows that the commutative, associative, and distributive laws hold for complex numbers.

THEOREM 5.1.   Let z_1, z_2 and z_3 be complex numbers. Then

(a) $z_1 + z_2 = z_2 + z_1$
(b) $z_1 + (z_2 + z_4) = (z_1 + z_2) + z_3$
(c) $z_1 z_2 = z_2 z_1$
(d) $z_1 (z_2 z_3) = (z_1 z_2) z_3$
(e) $z_1 (z_2 + z_3) = z_1 z_2 + z_1 z_3$

The **zero** complex number, denoted by 0, is defined by $0 = 0 + (0)i$ and the **unit** complex number is the number $1 + (0)i$.

THEOREM 5.2   For every complex number z,
(a) $0 + z = z + 0 = z$
(b) $[1 + (0)i]z = z[1 + (0)i] = z$

We shall prove part (b) and leave part (a) for the student to prove as an exercise.

Proof of (b). Let $z = x + yi$. Then by Definition 5.2, we have

$$
\begin{aligned}
[1 + (0)i]z &= [1 + (0)i][x + y^i] \\
&= [(1)x - (0)y] + [(1)y + (0)x]i \\
&= x + yi \\
&= z
\end{aligned}
$$

Similarly,

$$
\begin{aligned}
z[1 + (0)i] &= [x + yi][1 + (0)i] \\
&= [x(1) - y(0)] + [x(0) + y(1)]i \\
&= x + yi \\
&= z
\end{aligned}
$$

Therefore,

$$[1 + (0)i]z = z[1 + (0)i] = z$$

This completes the proof.

ANSWERS:

There is a one-to-one correspondence between the set of real numbers x and the set of complex numbers $z = x + (0)i$. Let

$$z_1 = x_1 + (0)i \quad \text{and} \quad z_2 = x_2 + (0)i$$

be two complex numbers. We find that

$$z_1 + z_2 = [x_1 + (0)i] + [x_2 + (0)i] = (x_1 + x_2) + (0)i$$

and

$$z_1 z_2 = [x_1 + (0)i][x_2 + (0)i] = x_1 x_2 + (0)i$$

Hence, complex numbers of the form $x + 0i$ behave like real numbers with respect to addition and multiplication. Because of this, we identify the real number x with the complex number $x + (0)i$, and we write

$$x = x + (0)i$$

Earlier we introduced the number $i = \sqrt{-1}$. We find that

$$(0 + i)(0 + i) = -1 + (0)i = -1$$

Thus, if we set $i = 0 + (1)i$, then $i^2 = -1$ as was indicated earlier.

If k is a real number we find that

$$k(a + bi) = [k + (0)i](a + bi)$$
$$= [ka - (0)b] + [kb + (0)a]i$$
$$= ka + kbi$$

For example, if $z = -1 + 5i$ then

$$4(-1 + 5i) = -4 + 20i$$

Self-test. Let $z = 2 - 3i$. Then

$-3z = $ _____

It should be noted that the Cartesian product may be used to define complex numbers. In this way the set C of complex numbers is defined by

$$C = R \times R$$

Thus, if $z \in C$, then z is an ordered pair of real numbers (x, y) and we write

$$z = (x, y)$$

and if $z_1 = (x_2, y_2)$ and $z_2 = (x_2, y_2)$ are two complex numbers, (It will be beneficial to review Definitions 5.1 and 5.2 for understanding.) then,

(a) $z_1 = z_2$ if and only if $x_1 = x_2$ and $y_1 = y_2$
(b) $z_1 + z_2 = (x_1, y_1) + (x_2, y_2) = (x_1 + x_2, y_1 + y_2)$
(c) $z_1 z_2 = (x_1, y_1)(x_2, y_2) = (x_1 x_2 - y_1 y_2, x_2 y_1 + x_1 y_2)$
(d) $kz_1 = k(x_1, y_1) = (kx_1, ky_1)$ where k is a real number.

ANSWERS: $!6 + 9-$

With this notation we identify the real number x with the complex number $(x,0)$ and we write

$$x = (x,0)$$

If we set $i = (0,1)$ then

$$i^2 = (0,1)(0,1) = (-1,0)$$

Thus $i^2 = -1$ as was indicated earlier.

If $z = (a,b)$ we can write

$$z = (a,b) = (a,0) + (0,b)$$

But

$$(0,b) = b(0,1) = bi$$

Therefore,

$$z = (a,0) + bi = a + bi$$

Indeed, there is a one-to-one correspondence between the symbols $a + bi$ and the ordered pair (a,b).

Clearly, if $z = a + bi$, then the **additive inverse** is $-z = (-1)z$ since

$$z + (-z) = (a+bi) - (a+bi) = (a-a) + (b-b)i = 0 + (0)i = 0$$

Conversely, if $z + w = 0$, then $w = -z$.

Self-test. Let $z = -1 + 2i$. Then

(a) $-z =$ _____ (b) $z + (-z) =$ _____

A more surprising result is the **multiplicative inverse** which is contained in the following theorem.

THEOREM 5.3 If $z = a + bi$, and $a^2 + b^2 \neq 0$, then z has a **multiplicative inverse** (or **reciprocal**) denoted by z^{-1} (or $1/z$) such that

$$zz^{-1} = 1 + (0)i = 1$$

Proof. Let

$$z^{-1} = \frac{a}{a^2+b^2} - \frac{b}{a^2+b^2} i$$

Using Definition 5.2, we find the following.

ANSWERS:

$$zz^{-1} = (a+ib)\left[\frac{a}{a^2+b^2} - \frac{b}{a^2+b^2}\,i\right]$$

$$= \frac{a^2}{a^2+b^2} + \frac{abi}{a^2+b^2} - \frac{abi}{a^2+b^2} - \frac{b^2i^2}{a^2+b^2}$$

$$= \frac{a^2}{a^2+b^2} + \frac{b^2}{a^2+b^2} + i\left(\frac{ab}{a^2+b^2} - \frac{ab}{a^2+b^2}\right)$$

$$= \frac{a^2+b^2}{a^2+b^2} + i(0) = 1$$

This completes the proof.

* **Self-test.** Let $z = 3 + 4i$. Then

    (a) $z^{-1} = $ _____ (b) $zz^{-1} = $ _____

We can now define division involving two complex numbers.

DEFINITION 5.3 If $w \neq 0$, the quotient z/w is defined by

$$\frac{z}{w} = z\left(\frac{1}{w}\right) = zw^{-1}$$

That is, to divide z by w, we merely multiply z by the reciprocal of w.

EXAMPLE 2. Compute $(2-7i)/(2+3i)$.

Solution. Here $w = 2 + 3i$

$$w^{-1} = [2/(2^2+3^2)] - [3i/(2^2+3^2)] = \tfrac{1}{13}(2-3i)$$

Therefore,

$$(2-7i)/(2+3i) = (2-7i)(\tfrac{1}{13})(2-3i)$$

$$= \tfrac{1}{13}(4 - 14i - 6i + 21i)^2$$

$$= \tfrac{1}{13}(4 - 21 - 20i) = \tfrac{1}{13}(-17 - 20i)$$

$$= -\tfrac{17}{13} - \tfrac{20}{13}i$$

** **Self-test.** Let $z = 1 + 2i$ and $w = 1 - 3i$. Then

    (a) $w^{-1} = $ _____ (b) $z/w = $ _____

NOTE: A valid question is: How do we obtain z^{-1}? To answer this question, let $z^{-1} = u + iv$. Then, $(a+bi)(u+iv) = 1$ or, $(au-bv) + i(av+bu) = (1,0)$ and by Definition 5.1: $au-bv = 1$; $av+bu = 0$. Solving for u and v, we obtain: $u = [a/(a^2+b^2)]$ and $v = [-b/(a^2+b^2)]$.

ANSWERS:

EXERCISES 5.1

In Exercises 1-45, perform each indicated operation and write the resulting complex number in the form $a+bi$.

1. $(1+3i)+(2+4i)$

2. $(5+2i)+(-6+3i)$

3. $(6+3i)+(-5+i)$

4. $(2,3)+(-2,3)$

5. $(2+3i)-(-4-i)$

6. $(8,7)-(6,-5)$

7. $(5,-6)-(4,1)$

8. $(7-5i)-(4+2i)$

9. $5i-(2+3i)$

10. $3(-2+i)$

11. $-\frac{1}{4}(0,3)$

12. $(4,1)(2,-3)$

13. $(1+2i)(1+4i)$

14. $(1-2i)(1+3i)$

15. $(1-2i)(1+2i)$

16. $5i(2-3i)$

17. $2i(5+i)$

18. $(1+2i)(4-3i)$

19. $(5-2i)(5+2i)$

20. $(2+i)(2-i)$

21. $(2+3i)^2$

22. $2i(1+i)^2$

23. $(1-i)^3$

24. $(1+i)^4$

25. $(2+3i)/(1+i)$

26. i^3

27. $1/i$

28. i^{-2}

29. $(1+i)/i$

30. $i/(1-i)$

31. $(\sqrt{2}+i)/(\sqrt{2}-i)$

32. $2/[5(1+i)]$

33. $3/(2+3i)$

34. $(3+4i)/(1-2i)$

35. $(4+6i)/(1-5i)$

36. $(6-7i)/(-1-5i)$

37. $4i^{-9}$

38. $(1+i)/(1-i)^2$

39. $(1-i)/(1+i)^2$

40. $2i^7$

41. $i^{16}-2i^5$

42. $[(1-3i)/(1+i)]+[(3+i)/(1-i)]$

43. $[(2-i)/(1+i)]+[(3-i)/(1+i)]$ 44. $[(2-6i)/(3+i)]-[(1-2i)/i]$

45. $3+[(2-i)/(1+3i)]$

46. Prove Theorem 5.1.

47. Prove part a of Theorem 5.2.

48. Prove that if z_1 and z_2 are complex numbers and $z_1z_2=0$, then either $z_1=0$ or $z_2=0$.

49. Prove that if z_1, z_2, and w are complex numbers, $z_1w=z_2w$, and $w\neq0$, then $z_1=z_2$.

In Exercises 50-52 find real numbers x and y satisfying each equation.

50. $x+2+3yi=12i$

51. $3x+4-2iy-i=-5+4i$

52. $2y+x-3xi=4+6i$

53. Show that each of $z_1=-1+i$ and $z_2=-1-i$ satisfies the equation $z^2+2z+2=0$.

54. Determine whether $1+2i$ is a solution of $z^2-2z+5=0$.

55. Determine whether $1+\sqrt{2}i$ is a solution of $z^2-2z+3=0$.

56. Determine whether $1-\sqrt{2}i$ is a solution of $z^2-2z+3=0$

57. Let n be a positive integer and compute each of the following.

   (a) i^{4n} (b) i^{4n+1} (c) i^{4n+2}

5.2 THE GEOMETRY OF COMPLEX NUMBERS

OBJECTIVES

1. *Set up a geometric representation of complex numbers.*
2. *Define and establish some properties of the conjugate and absolute value of a complex number.*

The geometric representation of the complex numbers first appeared in 1685 in the writings of the English mathematician John Wallis, and in 1797 the complex plane was used explicitly by the Norwegian surveyor Caspar Wessel (1745-1818). However, complex numbers remained unappreciated until the French mathematician Jean Robert Argand (1768-1822) rediscovered their geometric representation. Hence, the complex plane is often called the **Argand plane**.

We have seen that each complex number $z = x + iy$ can be associated with the ordered pair of real numbers (x,y) and each ordered pair of real numbers (x,y) can be associated with the complex number $z = x + iy$. Thus, there is a one-to-one correspondence between the set of ordered pairs of real numbers (x,y) and the set of complex numbers $z = x + iy$. Because of this, we can represent a complex number as a point in a plane. The plane on which the complex numbers are represented is called the **complex plane** (or **Argand plane**). The complex plane is constructed by taking the horizontal axis (x-axis) as the **real axis** and the vertical axis (y-axis) as the **imaginary axis** (see Figure 1). Geometric representation of the complex numbers $z_1 = 2 + 3i$, $z_2 = -2 + 2i$, and $z_3 = -3 - 2i$ is shown in Figure 1. Complex numbers of the form $z = x$ are represented by points of the form $(x,0)$ on the real axis and complex numbers of the form $z = yi$ are represented by points of the form $(0,y)$ on the imaginary axis.

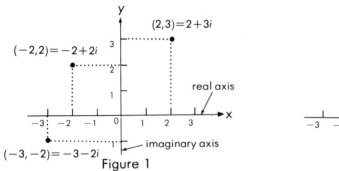

Figure 1

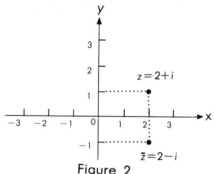

Figure 2

We notice that if $z=2+i$, then its mirror image in the real axis is the complex number $2-i$ (see Figure 2). We call this number the **conjugate** of z. In general, if $z=x+iy$, then the conjugate of z, denoted by $\bar{z}$, is given by $\bar{z}=\overline{x+iy}=x-iy$.

Some properties of the conjugate of a complex number are given in the following theorem.

THEOREM 5.4   If $z_1=x_1+iy_1$ and $z_2=x_2+iy_2$, then:

(a) $\overline{z_1+z_2}=\bar{z}_1+\bar{z}_2$

(b) $\overline{z_1z_2}=\bar{z}_1\bar{z}_2$

(c) $\overline{(z_1/z_2)}=(\bar{z}_1/\bar{z}_2)$,   provided $z_2\neq0$

(d) $\overline{(\bar{z}_1)}=z_1$

We shall prove Part (b) and leave the remaining parts for the student to prove as an exercise.

Proof.

$$\overline{z_1z_2}=\overline{(x_1+iy_1)(x_2+iy_2)}$$
$$=\overline{[(x_1x_2-y_1y_2)+i(x_1y_2+x_2y_1)]}$$
$$=(x_1x_2-y_1y_2)-i(x_1y_2+x_2y_1) \tag{1}$$

Also,

$$\bar{z}_1\bar{z}_2=\overline{(x_1+iy_1)}\,\overline{(x_2+iy_2)}$$
$$=(x_1-iy_1)(x_2-iy_2)$$
$$=(x_1x_2-y_1y_2)-i(x_1y_2+x_2y_1) \tag{2}$$

Therefore, from Equations 1 and 2, we have

$$\overline{z_1z_2}=\bar{z}_1\bar{z}_2$$

as asserted.

The student will recall that the absolute value of a real number x is represented by the length of the line segment from the origin to the point representing the number x on the real number line. We shall introduce a similar definition for the absolute value of a complex number.

DEFINITION 5.4   Let $z=(a,b)=a+ib$ be a complex number. Then the *ab-solute value* (or *modulus* or *magnitude*) of z, denoted by $|z|$, is defined by

$$|z|=\sqrt{a^2+b^2}$$

Clearly, the length of the line segment from the origin to the point representing z is given by $\sqrt{a^2+b^2}$ (see Figure 3).

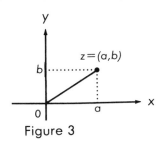

Figure 3

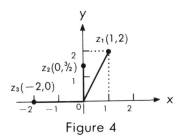

Figure 4

EXAMPLE 1. Compute the absolute value of the following.
   (a) $z_1 = 1+2i$
   (b) $z_2 = 3i/2$
   (c) $z_3 = -2$

Solution. We use Definition 5.4 (see Figure 4).
   (a) $|z_1| = |1+2i| = \sqrt{1^2+2^2} = \sqrt{5}$
   (b) $|z_2| = |3i/2| = \sqrt{0^2+(3/2)^2} = 3/2$
   (c) $|z_3| = |-2| = \sqrt{(-2)^2+0^2} = \sqrt{(-2)^2}=2$

Notice that the answer in Part c agrees with our definition of absolute values for real numbers. Also note that the absolute value of a complex number is a non-negative real number.

Self-test. Given that $z=3-i$, compute the following.
   (a) $\bar{z}$ =   _____
   (b) $|z|$ =   _____
   (c) $z\bar{z}$ =   _____
   (d) $z+\bar{z}$ =   _____

It should be noted that in this problem, $|z|^2=(\sqrt{10})^2=10=z\bar{z}$. In fact (see Problem 10 in Exercises 5.2), if z is any complex number, then

$$|z|^2=z\bar{z}$$

Other properties involving the absolute value of complex numbers are stated in the following theorem.

THEOREM 5.5 If z_1 and z_2 are complex numbers, then:

(a) $|z_1| \geqq 0$

(b) $|z_1 z_2| = |z_1||z_2|$

(c) $|\bar{z_1}| = |z_1|$

(d) $|-z_1| = |z_1|$

(e) $|z_1| \neq 0$ if and only if $z_1 = 0$

(f) $|z_1/z_2| = |z_1|/|z_2|$, provided $z_2 \neq 0$.

In Figure 5 we show the classification of numbers.

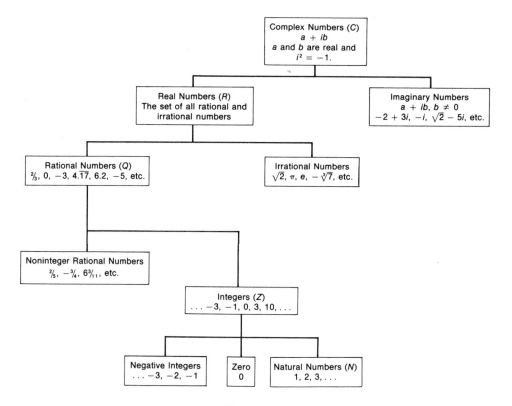

Figure 5

EXERCISES 5.2

In Exercises 1-8, evaluate the expression given that $z_1 = 2 - 3i$ and $z_2 = -1 + 4i$.

1. $\overline{z_1 + z_2}$ 2. $\overline{z_1} + \overline{z_2}$ 3. $z_1 + \overline{z_2}$ 4. $\overline{iz_1}$

5. $\overline{z_1 z_2}$ 6. $\overline{z_1}/\overline{z_2}$ 7. $\overline{z_1}\overline{z_2}$ 8. $\overline{(\overline{z_1})}$

9. Prove Parts a, c, and d of Theorem 5.4.

10. Show that for any complex number z, $|z|^2 = z\overline{z}$.

11. Prove Theorem 5.5.

In Exercises 12-17, compute the absolute value of each complex number.

12. $3 - 2i$ 13. $i(1 - 4i)$ 14. $2i/3$

15. $i/(1+i)$ 16. $(2+2i)/i$ 17. $(1+2i)\overline{(1-2i)}$

In Exercises 18-21, compute each quotient by using $\dfrac{z_1}{z_2} = \dfrac{z_1 \overline{z_2}}{z_2 \overline{z_2}} = \dfrac{z_1 \overline{z_2}}{|z_2|^2}$

18. $2/(3 - 2i)$ 19. $(2+i)/(2-i)$

20. $(5+9i)/(1-2i)$ 21. $(4+3i)/(3+2i)$

In Exercises 22-30, express the given number in the form $x + iy$.

22. $3 + (4+i)/(1-2i)$ 23. $6i + 1/(5-i)$ 24. $1/(1-i)^3$

25. $(1+i)^3/(1-i)$ 26. $(4-i^2)/(i-2)$ 27. $(25 - 4i^2)/(5+2i)$

28. $(2-3i)/(1+i)+(4+2i)/(3+i)$ 29. $(1-4i)/(2-i)+(7+3i)/(1+2i)$

30. $(1+i)^3/(1+i) - (2-i)/(3+2i)$

31. Show that if z satisfies the equation $az^2 + bz + c = 0$ where, a, b, and c are real numbers, then $\overline{z}$ also satisfies the equation.

32. If z_1 and z_2 are complex numbers, show that

$$\text{Re}\left(\frac{z_1}{z_1+z_2}\right) + \text{Re}\left(\frac{z_2}{z_1+z_2}\right) = 1$$

33. If z is a complex number, prove each of the following.
(a) $z = \overline{z}$ if and only if z is real.
(b) $z - \overline{z}$ is an imaginary number.
(c) $z + \overline{z}$ is a real number.

5.3 POLAR COORDINATES

OBJECTIVES

1. Set up the polar coordinate system.
2. Given the rectangular coordinates of a point find its polar coordinates.
3. Given the polar coordinates of a point find its rectangular coordinates.
4. Graph some simple polar equations.

So far, we have exclusively used the rectangular coordinate system for locating points in the plane. Besides the rectangular, there are other systems for represen-ting points and curves analytically. Of these others, probably the most important is the **polar coordinate system.** In this system, we start with a fixed point O called the pole or **origin.** From the pole a fixed ray (half-line) OA is drawn. The fixed line OA is called the **polar axis (or polar line).** The polar axis is usually drawn horizon-tally and to the right as shown in Figure 1.

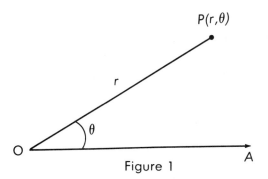

Figure 1

Now, let P be any point in the plane distinct from O. Let r be the distance from O to P and let θ be the radian measure of the angle AOP, having OA as the initial side and OP as the terminal side. Then the numbers r and θ serve as the polar coor-dinates for the point P and we write these coordinates as (r,θ). With each pair of numbers (r,θ) we associate a point P as follows.

By rotating the polar line OA through an angle $|\theta|$, counterclockwise if θ is positive and clockwise if θ is negative, we obtain ray OM. Then on ray OM, a point P is located so that $d(O,P)=r$ if $r\geq0$. If $r<0$, then the ray is extended backwards through O, and P is located on this extension so that $d(O,P)=|r|$. For example (see Figure 2), the point $P(2,\pi/3)$ is determined by first rotating counterclockwise the line OA, so that the radian measure of the angle AOM is $\pi/3$. Then the point on the line OM, which is two units from the pole O, is the point P. Another set of polar coordinates for the same point P is $(2,-5\pi/3)$. Furthermore, the polar coordinates $(2,7\pi/3)$ would also yield the same point. In general, the coordinates $(2,\pi/3+2n\pi)$, where n is any integer, give the same point as $(2,\pi/3)$. Still other sets of polar coordinates for the point $P(2,\pi/3)$ are $(-2,4\pi/3)$ and $(-2,-2\pi/3)$. Thus, if P has polar coordinates (r,θ), then P also has polar coordinates $(r,\theta+2n\pi)$ and $(-r,\theta+(2n-1)\pi)$ for every integer n.

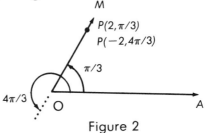

Figure 2

It is clear that to each pair of numbers (r,θ) there corresponds a **unique** point P in the plane. However, if we are given a point P in the plane, there is an unlimited number of polar coordinates that correspond to P. This is unlike the rectangular (Cartesian) coordinate system where there is a one-to-one correspondence between the rectangular coordinates and the points in the plane. Another example to show the nonuniqueness of the polar coordinates of a given point is obtained by considering the polar coordinates of the pole. If P is at the pole, then $r=0$, but θ can be any real number. Thus, the polar coordinates of the pole are $(0,\theta)$. Unique polar coordinates for a point P, not the pole, are sometimes obtained by restricting r and θ so that $r>0$ and $0\leq\theta<2\pi$.

EXAMPLE 1. Plot the point $P(3,-\pi/4)$. Find the other sets of polar coordinates (r,θ) for the same point P with r negative and $-2\pi<\theta<2\pi$.

Solution. The point P is plotted by drawing the angle of radian measure $\pi/4$ in a clockwise direction from the polar axis. Since $r>0$, P is on the terminal side of the angle, three units from the pole (see Figure 3a). Other sets of polar coordinates for P are $(-3,3\pi/4)$ and $(-3,-5\pi/4)$, illustrated in Figures 3b and 3c.

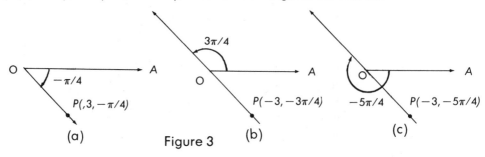

(a)   Figure 3   (b)   (c)

Self-test. Plot each of the following points on polar graph paper.

   (a) P(4,30°)   (b) Q(-2,7π/3)   (c) R(-1.5,-π/3)

We often wish to refer to both the rectangular and the polar coordinates of a point. To do this, we superimpose a rectangular coordinate system on the polar coordinate system, so that the origins in both systems coincide and the positive x-axis coincides with the polar line, and the ray for which $\theta = \pi/2$ is the positive y-axis (see Figure 4). Then, each point P has two types of coordinates, a rectangular set (x,y) and a polar set (r,θ).

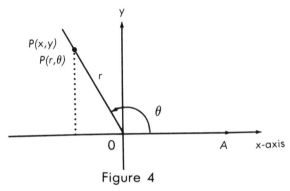

Figure 4

It is easy to see from Figure 4, that

$$x = r \cos \theta \tag{1}$$
$$y = r \sin \theta \tag{2}$$

From Equations 1 and 2 , we can obtain the rectangular coordinates of a point when its polar coordinates are known. To obtain equations which give a set of polar coordinates of a point when its rectangular coordinates are known, we square both sides of Equations 1 and 2 and add to obtain

$$x^2 + y^2 = r^2\cos^2\theta + r^2\sin^2\theta$$
$$= r^2(\cos^2\theta + \sin^2\theta)$$
$$= r^2$$

which gives

$$r = \pm\sqrt{x^2 + y^2}. \tag{3}$$

Dividing Equation 2 by Equation 1, we get

$$\frac{y}{x} = \frac{r \sin \theta}{r \cos \theta}$$

or

$$\frac{y}{x} = \tan \theta \tag{4}$$

Using Equations 3 and 4, we can obtain polar coordinates from rectangular coordinates.

EXAMPLE 2. Plot the point whose polar coordinates are $(-3, -4\pi/3)$, and find its rectangular coordinates.

Solution. The point is plotted in Figure 5.

$$x = r\cos\theta \qquad\qquad y = r\sin\theta$$
$$= -3\cos(-4\pi/3) \qquad = -3\sin(-4\pi/3)$$
$$= -3\cos(2\pi/3) \qquad = -3\sin 2\pi/3$$
$$= 3\cos\pi/3 \qquad\qquad = -3\sin\pi/3$$
$$= 3/2 \qquad\qquad\qquad = -3\sqrt{3}/2$$

Thus, the rectangular coordinates of the point P are $(3/2, -3\sqrt{3}/2)$.

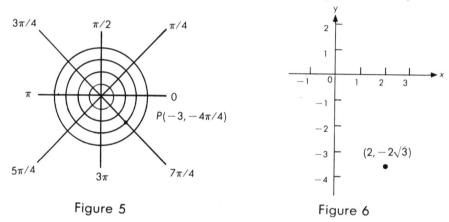

Figure 5   Figure 6

EXAMPLE 3. The rectangular coordinates of a point are $(2, -2\sqrt{3})$. Find all possible sets of polar coordinates for this point.

Solution. The point $(2, -2\sqrt{3})$ is plotted in Figure 6. For $r > 0$ and Equation 3, we get

$$r = \sqrt{(2)^2 + (-2\sqrt{3})^2} = 4$$

From Equation 4, $\tan\theta = -2\sqrt{3}/2 = -\sqrt{3}$, and since the point P is in the fourth quadrant we have a value of $\theta = -\pi/3$ as one of the solutions of $\tan\theta = -\sqrt{3}$. Thus, $(4, -\pi/3)$ is one choice of the polar coordinates for P. From our discussion in this section we conclude that all polar coordinates of P are given by

$$(4, -\pi/3 + 2n\pi) \text{ and } (-4, -\pi/3 + (2n-1)\pi)$$

where n is any integer.

Self-test.

(a) Change $P(-3, 60°)$ to rectangular coordinates: _____

(b) Change $P(1, \sqrt{3})$ into polar form with $r \geq 0$: _____

(c) In part (b) if $r < 0$, then the polar coordinates of P are: _____

Just as in the rectangular coordinate plane as an equation in x and y has a graph in the Cartesian plane, similarly an equation in r and θ has a graph in the polar coordinate plane. Thus, the graph of an equation r and θ consists of those and only those points P having some pair of coordinates satisfying the given equation. If an equation of a graph is given in polar coordinates, it is called a polar equation. To sketch the graph of a polar equation $F(r,\theta)=0$, we select a sequence of values for θ and compute the associated values for r. The desired graph is then obtained by plotting the ordered pairs (r,θ) and connecting them with a smooth curve. We now consider some examples of graphs in polar coordinates.

EXAMPLE 4. Graph $r=k$ (for all θ) where k is a positive constant.

Solution. The graph of the equation $r=k$ is a circle with the center at the pole and radius $r=k$. See Figure 7.

It should be noted that the graph of $r=-k$ is the same circle as above.

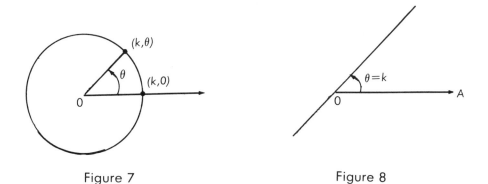

Figure 7   Figure 8

EXAMPLE 5. Graph $\theta=k$ where k is a constant.

Solution. See Figure 8. The graph of the equation $\theta=k$ is a straight line passing through the pole and making an angle of radian measure k with the polar axis.

It should be noted that the graph of $\theta=k+n\pi$, where n is any integer is the same as the line above.

EXAMPLE 6. Sketch the graph of $r=\sin\theta$.

Solution. To sketch a graph of a given equation we choose some values for θ and find the corresponding values of r as in Table 1.

θ	0	π/6	π/4	π/3	π/2	2π/3	3π/4	5π/6	π
r	0	½	$\sqrt{2}/2$	$\sqrt{3}/2$	1	$\sqrt{3}/2$	$\sqrt{2}/2$	½	0

Table 1

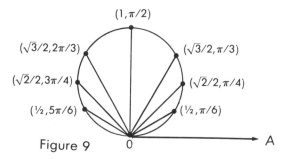

Figure 9   0   A

We note that as θ increases from 0 to $\pi/2$, r takes on values between 0 and 1. As θ increases from $\pi/2$ to π, r decreases from 1 to 0. We plot the points in the table and connect them with a smooth curve as in Figure 9. This is a circle with the center at $(1/2, \pi/2)$ and the radius $1/2$.

It can be easily verified that as θ increases from π to 2π, the circle is retraced. In fact, since $\sin(\theta + 2n\pi) = \sin\theta$ where n is an integer, the circle will continue to be retraced for $\theta > 2\pi$.

EXAMPLE 7. Sketch a graph of $r = 2(1 + \cos\theta)$.

Solution. We start by setting up a table (see Table 2) of values for θ with the corresponding values for r.

θ	0	$\pi/6$	$\pi/4$	$\pi/3$	$\pi/2$	$2\pi/3$	$3\pi/4$	$5\pi/6$	π
r	4	3.73	3.41	3	2	1	0.59	0.27	0

Table 2

We note that as θ increases from 0 to $\pi/2$, $\cos\theta$ decreases from 1 to 0 and r changes from 4 to 2; and as θ changes from $\pi/2$ to π, r changes from 2 to 0. Also, as θ changes from π to 2π, r changes from 0 to 4. (See Table 3.)

θ	π	$7\pi/6$	$5\pi/4$	$4\pi/3$	$3\pi/2$	$5\pi/3$	$7\pi/4$	$11\pi/3$	2π
r	0	0.27	0.59	1	2	3	3.41	3.73	4

Table 3

The graph of the equation $r = 2(1 + \cos\theta)$ is sketched in Figure 10. This heart shaped curve is called a **cardioid**.

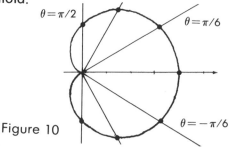

$\theta = \pi/2$

$\theta = \pi/6$

$\theta = -\pi/6$

Figure 10

EXAMPLE 8. Sketch a graph of $r = 3 \sin 2\theta$.

Solution. We set up a table of values for θ with the corresponding values for r (see Table 4). Note that in this case, we let 2θ take on the set of values $\pi/6$, $\pi/4$, $\pi/3$ and the corresponding angle values in the other quadrants.

θ	0	$\pi/12$	$\pi/8$	$\pi/4$	$\pi/3$	$3\pi/8$	$\pi/2$	$5\pi/8$	$3\pi/4$	$7\pi/8$	π
2θ	0	$\pi/6$	$\pi/4$	$\pi/2$	$2\pi/3$	$3\pi/4$	π	$5\pi/4$	$3\pi/2$	$7\pi/4$	2π
r	0	$3/2$	$3\sqrt{2}/2$	3	$3\sqrt{3}/3$	$3\sqrt{2}/2$	0	$-3\sqrt{2}/2$	-3	$-3\sqrt{2}/2$	0

Table 4

We note that for $\pi \le \theta < 2\pi$, $\sin 2\theta$ repeats the values it took on $0 \le \theta < \pi$. Thus plotting these values of r and θ and connecting them with a smooth curve we obtain the graph shown in Figure 11.

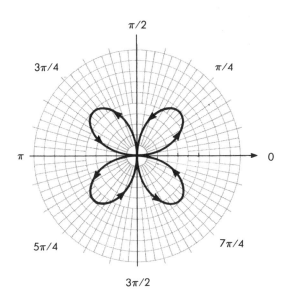

Figure 11

The curve is called a four-leafed rose. In general, the graph of an equation of the form

$$r = a \sin n\theta \quad \text{or} \quad r = a \cos n\theta$$

where a is a constant and n is an integer, is a rose having n leaves if n is odd and $2n$ leaves if n is even.

EXERCISES 5.3

In Exercises 1-12, use polar graph paper and locate the points which have the given polar coordinates.

1. $(2, \pi/3)$ 2. $(3, 3\pi/4)$ 3. $(-2, 5\pi/6)$

4. $(4, -30°)$ 5. $(2, -2\pi/3)$ 6. $(2, -9\pi/4)$

7. $(-3, -270°)$ 8. $(-1, -5\pi/4)$ 9. $(-5, -\pi/4)$

10. $(-\sqrt{2}, -4\pi/3)$ 11. $(\sqrt{2}, -\pi/3)$ 12. $(-4, 11\pi/4)$

13. Find the rectangular coordinates of each of the points in Exercises 1-12.

In Exercises 14-21, the rectangular coordinates of a point are given. Find a set of polar coordinates for each of the points.

14. $(1,1)$ 15. $(1, \sqrt{3})$ 16. $(1, -\sqrt{3})$

17. $(0,2)$ 18. $(2\sqrt{3}, 2)$ 19. $(-4,0)$

20. $(\sqrt{3}, -\sqrt{3})$ 21. $(-\sqrt{3}, -\sqrt{3})$

In Exercises 22-27, convert the rectangular equations into equations in polar coordinates.

22. $3x + 2y = 4$ 23. $xy = 1$ 24. $x^2 + y^2 = 9$

25. $x^2 + y^2 - 2x = 0$ 26. $x^2 = 16y$ 27. $y^2 = 4x$

In Exercises 28-32, convert the polar equation into equations in the rectangular coordinates.

28. $r = 4$ 29. $r = 1 + \cos \theta$ 30. $r = \sin \theta$

31. $r \cos \theta = 4$ 32. $r(1 - \cos \theta) = 1$

33. Show that the distance between the two points $P(r_1, \theta_1)$ and $Q(r_2, \theta_2)$ is given by

$$d(P,Q) = \sqrt{r_1^2 + r_2^2 - 2r_1 r_2 \cos(\theta_2 - \theta_1)}$$

In Exercises 34-47, sketch the graph of the given polar equations.

34. $r=3$ 35. $\theta=\pi/3$

36. $r=4\cos\theta$ 37. $r=1+\sin\theta$

38. $r\cos\theta=2$ 39. $r\sin\theta=-1$

40. $r=2(1-\cos\theta)$ 41. $r=2(1-\sin\theta)$

42. $r=\theta$ 43. $r=\cos2\theta$

44. $r=2\cos3\theta$ 45. $r\theta=\pi, \theta>0$

46. $r=\theta/\pi, \theta\geq0$ 47. $r^2\theta=9$

48. Find the points of intersection of
$$r=1+\cos\theta \quad\text{and}\quad r=3\cos\theta$$

49. Find the points of intersection of
$$r=2 \quad\text{and}\quad r=2(1-\sin\theta)$$

50. Sketch the graph of the equation
$$r=\frac{4e}{1-e\cos\theta}$$
for each of the values of e and identify the curve.
(a) $e=1$ (b) $e=\frac12$ (c) $e=2$

5.4 POLAR FORM OF COMPLEX NUMBERS

OBJECTIVES

1. Write a complex number in polar form.
2. Use polar form to multiply and divide complex numbers.
3. Use de Moivre's formula to find powers of complex numbers.
4. Find all the nth roots of a complex number.

In Section 5.2 we noted that each complex number $z=x+iy$ may be associated with a point (x,y) in the complex plane. Thus, the point whose rectangular coordinates are $(2,4)$ represents the complex number $z=2+4i$. Recall that a complex plane is a coordinate plane with a complex number assigned to each point.

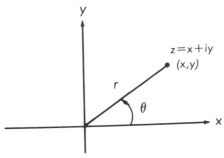

Figure 1

In Figure 1, we indicate the geometric representation of the complex number $z=x+iy$. This representation leads to a polar representation for a complex number. Referring to Figure 1, the nonzero complex number $z=x+iy$ can be located with the polar coordinates (r,θ) where

$$r=|z|=\sqrt{x^2+y^2} \tag{1}$$

and

$$x=r\cos\theta, \quad y=r\sin\theta \tag{2}$$

Hence the complex number $z=x+iy$ can be written in the **polar form** (or **trigonometric form**)

$$z=r \cos \theta + ir \sin \theta$$
$$=r(\cos \theta + i \sin \theta) \qquad (3)$$

The "cosine $+ i$ sine" prompts us to use the abbreviation

$$z=r \text{ cis } \theta$$

Now, suppose we have $z=1+i$. Then $r=\sqrt{1^2+1^2}=\sqrt{2}$, and clearly one value of θ is $\theta=\pi/4$. Thus

$$1+i=\sqrt{2}[\cos(\pi/4)+i \sin(\pi/4)] =\sqrt{2} \text{ cis}(\pi/4)$$

However, we may also write

$$1+i=\sqrt{2}[\cos(\pi/4+2k\pi)+i \sin(\pi/4+2k\pi)]$$

for $k=0,1,2,\ldots$ (see Figure 2) .

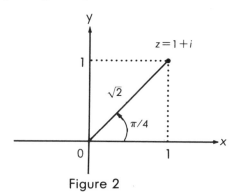

Figure 2

One should observe that whenever z is expressed in polar form with $r>0$, then r must be equal to $|z|$. However, there are an infinite number of values for the angle θ, each differing by an integer multiple of 2π. Thus the polar form for a complex number is not unique.

In the polar form for a complex number z, we call θ the **argument** (or **amplitude** or **angle**) of z, and we denote it by **arg z**. The value of arg z which lies in the interval $(-\pi, \pi]$ is referred to as the **principal value of the argument** and is denoted by **Arg z** (note the capitalization of A).

EXAMPLE 1. Find the polar form of the following complex numbers.

(a) $-1+\sqrt{3}i$ (b) $1-i$ (c) $2+5i$

Solution.

(a) For the complex number $z=-1+\sqrt{3}i$ (see Figure 3), we have $r=|-1+\sqrt{3}i| =\sqrt{(-1)^2+(\sqrt{3})^2}=2$ and $\sin\theta=\sqrt{3}/2$, $\cos\theta=-\frac{1}{2}$. Thus, Arg $z=\theta=2\pi/3$ (or $120°$), and we have

$$-1+\sqrt{3}i=2[\cos(2\pi/3)+i\sin(2\pi/3)]=2\text{ cis}(2\pi/3)$$

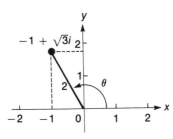

Figure 3

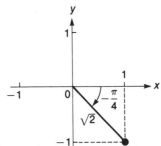

Figure 4

(b) For the complex number $z=1-i$ (see Figure 4), we have $r=|1-i|=\sqrt{1^2+(-1)^2} =\sqrt{2}$ and $\sin\theta=-\sqrt{2}/2$, $\cos\theta=\sqrt{2}/2$. Thus, Arg $z=\theta=-\pi/4$ (or $-45°$), and we have

$$1-i=\sqrt{2}[\cos(-\pi/4)+i\sin(-\pi/4)]=\sqrt{2}\text{ cis}(-\pi/4)$$

(c) For the complex number $z=2+5i$ (see Figure 5), we have $r=|2+5i|=\sqrt{2^2+5^2}=\sqrt{29}$ and Arg $z=\theta=\text{Arctan}(\frac{5}{2})$. Thus,

$$2+5i=\sqrt{29}[\cos(\text{Arctan }\tfrac{5}{2})+i\sin(\text{Arctan }\tfrac{5}{2})]=\sqrt{29}\text{ cis}(\text{Arctan }\tfrac{5}{2})$$

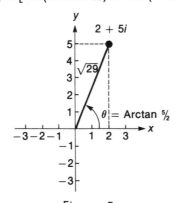

Figure 5

EXAMPLE 2. Find the rectangular form for the complex number 2 cis $135°$.

Solution. Here we have $z=2$ cis $135° =x+iy$ where

$$x=2\cos 135° =2(-\sqrt{2}/2)=-\sqrt{2}$$

and

$$y=2\sin 135° =2(\sqrt{2}/2)=\sqrt{2}$$

Thus

$$2\text{ cis }135° =-\sqrt{2}+\sqrt{2}i$$

Self-test. Given that $z = 4 - 4i$, fill in the blanks.

(a) $r = $ _____

(b) Arg $z = $ _____

(c) The polar form of z is: _____

The polar form is inconvenient for the addition of complex numbers. However, multiplication, division, and taking roots of complex numbers are somewhat easier to do with polar notation. If we let

$$z_1 = r_1(\cos\theta_1 + i\sin\theta_1)$$
$$z_2 = r_2(\cos\theta_2 + i\sin\theta_2)$$

then we compute $z_1 z_2$ as follows.

$$z_1 z_2 = r_1 r_2[(\cos\theta_1\cos\theta_2 - \sin\theta_1\sin\theta_2) + i(\sin\theta_1\cos\theta_2 + \cos\theta_1\sin\theta_2)]$$
$$= r_1 r_2[\cos(\theta_1 + \theta_2) + i\,(\sin(\theta_1 + \theta_2)]$$
$$= r_1 r_2\,\text{cis}(\theta_1 + \theta_2)$$

That is, the argument of the product is the sum of the arguments.

Thus, we have established the following:

THEOREM 5.6   If the polar form of two complex numbers z_1 and z_2 are given by

$$z_1 = r_1\,\text{cis}\,\theta_1 \quad\text{and}\quad z_2 = r_2\,\text{cis}\,\theta_2 \tag{4}$$

then

$$z_1 z_2 = r_1 r_2\,\text{cis}(\theta_1 + \theta_2) \tag{5}$$

To divide complex numbers we state the following theorem.

THEOREM 5.7   If the polar form for two complex numbers z_1 and z_2 is given by Equation 4, then

$$z_1/z_2 = (r_1/r_2)[\cos(\theta_1 - \theta_2) + i\,\sin(\theta_1 - \theta_2)] = (r_1/r_2)\,\text{cis}(\theta_1 - \theta_2) \tag{6}$$

provided $r_2 \neq 0$.

ANSWERS:

(a) $4\sqrt{2}$ (b) $-\pi/4$ (c) $4\sqrt{2}\,\text{cis}(-\pi/4)$

EXAMPLE 3. Given that $z_1 = -2\sqrt{3}+2i$ and $z_2 = 1-\sqrt{3}i$, use polar form to find each of the following.

$$\text{(a) } z_1 z_2 \qquad\qquad\qquad \text{(b) } z_1/z_2$$

Solution. First we find the polar forms for z_1 and z_2.

$$z_1 = -2\sqrt{3}+2i = 4 \text{ cis}(5\pi/6)$$
$$z_2 = 1-\sqrt{3}i = 2 \text{ cis}(-\pi/3)$$

(a) We use Theorem 5.6

$$z_1 z_2 = (4)(2)\text{cis } [5\pi/6+(-\pi/3)] = 8 \text{ cis}(\pi/2) = 8i$$

(b) We use Theorem 5.7

$$z_1/z_2 = (4/2) \text{ cis}[5\pi/6-(-\pi/3)]$$
$$= 2 \text{ cis}(7\pi/6)$$
$$= 2 \text{ cis}(-5\pi/6) = -\sqrt{3}-i$$

Self-test. Let $z_1 = 5 \text{ cis } 35°$ and $z_2 = 2 \text{ cis } 25°$.

(a) $z_1 z_2 = $ _____

(b) $z_1/z_2 = $ _____

The Product Formula 5 can easily be extended using mathematical induction to give the product of n factors.

$$z_1 z_2 ... z_n = r_1 r_2 ... r_n \text{cis}(\theta_1 + \theta_2 + ... \theta_n) \qquad (7)$$

A special case of Equation 7 is when $z_1 = z_2 = ... = z_n = z$. We find:

$$zz = z^2 = r^2 \text{ cis } 2\theta$$
$$z^2 z = r^2 r \text{ cis}(2\theta+\theta) \quad \text{or} \quad z^3 = r^3 \text{ cis } 3\theta$$
$$z^3 z = r^3 r \text{ cis}(3\theta+\theta) \quad \text{or} \quad z^4 = r^4 \text{ cis } 4\theta$$

Continuing this process, it appears that for every positive integer n

$$z^n = r^n \text{ cis } n\theta \qquad (8)$$

ANSWERS:

In fact, Formula 8 can be established for every positive integer n, using mathematical induction. It is also possible to show that Formula 8 is valid when n is a negative integer. The result is known as **de Moivre's formula**, named after the French mathematician Abraham de Moivre*. De Moivre proved this result for positive integers n. It was Euler who generalized the result to include all real n.

EXAMPLE 4. Compute each of the following.

(a) $(1+i)^{12}$ (b) $128/(\sqrt{3}-i)^9$

Solution.
(a) Writing $(1+i)$ in polar form, we have
$$1+i=\sqrt{2}\ cis(\pi/4)$$
We then use Formula 8.
$$(1+i)^{12}=(\sqrt{2})^{12}\ cis[12(\pi/4)]=64\ cis\ 3\pi=64\ cis\ \pi=-64$$

(b) We express $\sqrt{3}-i$ in polar form.
$$\sqrt{3}-i=2\ cis(-\pi/6)$$
We then use de Moivre's formula, with $n=-9$, $r=2$, and $\theta=-\pi/6$.
$$z=128(2)^{-9}\{\cos[(-9)(-\pi/6)]+i\ \sin[(-9)(-\pi/6)]\}$$
$$=128/512\ cis(3\pi/2)=\tfrac{1}{4}\ cis(-\pi/2)=-\tfrac{1}{4}i$$

Self-test. Compute the following and write your answer in rectangular form.

$[4\ cis(\pi/6)]^3 = $ _____ = _____.

As in the case of real numbers, a complex number w is said to be an **nth root** of the complex number z if

$$w^n=z \qquad\qquad\qquad (9)$$

---

*ABRAHAM DE MOIVRE

Abraham de Moivre (1667-1754) was born in the province of Champagne, France on May 20, 1667, moving to London in 1685. In London he studied mathematics and was acquainted with such other noted scientists of his time as Issac Newton and Edmond Halley. De Moivre became an accomplished mathematician and was elected to the London Royal Society in 1697.

De Moivre is considered to be one of the fathers of probability theory, a concept which he put to great practice in the many hours he spent in London's gambling houses. He even wrote a manual for gamblers called **Doctorine of Chances**. De Moivre died in 1754, having the dubious distinction of correctly predicting the day of his death.

ANSWERS:

64 cis(π/2)=64i

For example:

$$i^{1/2} = \sqrt{2}/2(1+i) \quad \text{since} \quad [\sqrt{2}/2(1+i)]^2 = i$$

and

$$i^{1/2} = -\sqrt{2}/2(1+i) \quad \text{since} \quad [-\sqrt{2}/2(1+i)]^2 = i$$

We shall now show that every nonzero complex number has n distinct nth roots. In Equation 9, we regard z as a given number and solve for the unknown w. Let z and w have the polar forms

$$z = r_0 \text{ cis } \theta_0 \quad \text{and} \quad w = r \text{ cis } \theta \tag{10}$$

Substituting Equations 10 into Equation 9 and applying de Moivre's formula, we obtain

$$r^n \text{cis } n\theta = r_0 \text{ cis } \theta_0 \tag{11}$$

Now, if two complex numbers are equal, so are their absolute values. Consequently, $r^n = r_0$, and since r and r_0 are positive, we get

$$r = \sqrt[n]{r_0} \tag{12}$$

Also from Equation 11, we have

$$\cos n\theta = \cos \theta_0 \quad \text{and} \quad \sin n\theta = \sin \theta_0 \tag{13}$$

The relations in Equation 13 are true if and only if $n\theta = \theta_0$ or $n\theta$ differs from θ_0 by a multiple of 2π. That is,

$$n\theta = \theta_0 + 2\pi k, \ k \in Z$$

or

$$\theta = (\theta_0 + 2k\pi)/n, \ k \in Z$$

Thus,

$$w_k = \sqrt[n]{r_0} \text{ cis}[(\theta_0 + 2k\pi)/n] \tag{14}$$

for some integer k. It may appear that we can generate infinitely many values for w. This, however, is not the case. If we substitute $k = 0, 1, 2, \ldots, n-1$ successively in Equation 14, there result n distinct values for w and hence n distinct nth roots of z. For other values of k, we merely duplicate these n roots. For example, when $k = n$, we get

$$\frac{\theta_0 + 2n\pi}{n} = (\theta/n) + 2\pi$$

which produces the same number as $k = 0$.

We have in effect proven the following theorem.

THEOREM 5.8 If $z = r$ cis θ is any nonzero complex number and if n is any positive integer, then z has exactly n distinct nth roots. These roots are given by

$$w_k = \sqrt[n]{r} \text{ cis}[(\theta + 2k\pi)/n] \tag{15}$$

where $k = 0,1,2,\ldots,n-1$.

We note that the nth roots of z all have absolute value $\sqrt[n]{r}$. Hence the points representing the complex numbers $w_0, w_1, w_2, \ldots, w_{n-1}$ all lie on a circle with its center at the origin and with the radius $\sqrt[n]{r}$. These points are equally spaced on the circle $2\pi/n$ radians apart.

EXAMPLE 5. Find all the fourth roots of $2(-1+\sqrt{3}i)$.

Solution. In polar form, we have

$$2(-1+\sqrt{3}i) = 4 \text{ cis } 120°$$

Using Equation 15 with $n = 4$, $\theta = 120°$, and $r = 4$, we obtain

$$w_k = \sqrt[4]{4} \text{ cis}[(120° + 360°k)/4] \qquad k = 0,1,2,3$$

Hence:

$$\begin{aligned}
w_0 &= \sqrt{2} \text{ cis } 30° &&= \sqrt{2}[(\sqrt{3}/2) + \tfrac{1}{2}i] \\
w_1 &= \sqrt{2} \text{ cis } 120° &&= \sqrt{2}[-\tfrac{1}{2} + (\sqrt{3}/2)i] \\
w_2 &= \sqrt{2} \text{ cis } 210° &&= \sqrt{2}[(-\sqrt{3}/2) - \tfrac{1}{2}i] \\
w_3 &= \sqrt{2} \text{ cis } 300° &&= \sqrt{2}[\tfrac{1}{2} - (\sqrt{3}/2i)]
\end{aligned}$$

The points representing $2(-1+\sqrt{3}i)$ and the fourth roots are shown in Figure 6.

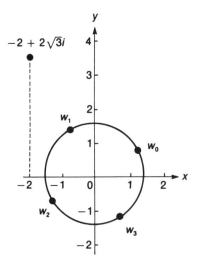

Figure 6

EXERCISES 5.4

In Exercises 1-12, change each complex number to polar form.

1. $-1+i$ 2. $-1-i$ 3. $1+\sqrt{3}i$

4. $-4+4i$ 5. $2\sqrt{3}-2i$ 6. $-10i$

7. $2i$ 8. -37 9. 17

10. $\frac{1}{2}(\sqrt{3}+i)$ 11. $-3(1+i)$ 12. $2i(\sqrt{3}+i)$

In Exercises 13-20, convert each number to polar form; then find $z_1 z_2$ and z_1/z_2.

13. $z_1=1+i$, $z_2=1-i$ 14. $z_1=\sqrt{3}+i$, $z_2=1+i$

15. $z_1=\sqrt{3}-i$, $z_2=5(\sqrt{3}+i)$ 16. $z_1=3i$, $z_2=-2\sqrt{3}+2i$

17. $z_1=2i$, $z_2=2\sqrt{3}-2i$ 18. $z_1=1+\sqrt{3}i$, $z_2=4-4i$

19. $z_1=i$, $z_2=-i$ 20. $z_1=\sqrt{3}-i$, $z_2=3-3\sqrt{3}i$

In Exercises 21-30, evaluate each expression by means of de Moivre's formula and express your answers in rectangular form.

21. $(1+i)^9$ 22. $(1-i)^5$

23. $(\sqrt{3}+i)^3$ 24. $(-1+\sqrt{3}i)^4$

25. $(-2+2i)^4$ 26. $[2\,\mathrm{cis}(\pi/3)]^3$

27. $[3\,\mathrm{cis}(\pi/3)]^4$ 28. $[2\,\mathrm{cis}(\pi/5)]^5$

29. $2/(1-i)^3$ 30. $4/(1-\sqrt{3}i)^3$

31. Find the two square roots of $1-\sqrt{3}i$.

32. Find the two square roots of $4i$.

33. Find the two square roots of $2+2\sqrt{3}i$.

34. Find all the cube roots of $8i$.

35. Find all the fourth roots of $8+8\sqrt{3}i$.

36. Find all the cube roots of -1.

37. Find all the fourth roots of -1.

38. Find all the fifth roots of $-\sqrt{3}+i$.

39. Prove Theorem 5.7.

40. Prove de Moivre's formula using mathematical induction.

41. Prove that de Moivre's formula is true for negative integers n.

REVIEW OF CHAPTER FIVE

1. The set of complex numbers C is defined by C= _____.

2. In the set of complex numbers C, define each of the following:
   (a) Equality   (b) Sum   (c) Difference
   (d) Product   (e) Multiplicative Inverse (f) Quotient

3. If z is a complex number, define each of the following and illustrate each geometrically.
   (a) Conjugate of z (b) Absolute value of z

In Exercises 4-10, write each expression in the form $x+iy$.

4. $(1-i)+(2+3i)$ 5. $(1-i)-(3-2i)$

6. $(2-2i)(4+3i)$ 7. $(3-2i)/(3-i)$

8. $-i^7$ 9. $1/(3-2i)$

10. $\overline{(2+i)}(1+3i)$

11. Solve for x and y: $3x + 2i = 9 - (3 + y)i$

12. Determine whether $-1 - 2i$ is a solution of $x^2 + 2x + 5 = 0$. Is $-1 + 2i$ a solution? $1 - 2i$?

13. Find $|3 - 4i|$

14. Show that for any complex number z, $|z| = |-z|$.

15. Show that for any complex number z, $|z| = |\bar{z}|$.

16. If $z = \sqrt{3} + i$, then:
    (a) Arg z = _____
    (b) Modulus z = _____

17. Plot each of the following points on polar graph paper.
    (a) $(3, 120°)$ (b) $(-5, -7\pi/3)$ (c) $(-3, 2\pi/3)$ (d) $(5, 7\pi/3)$

18. Find the rectangular coordinates of each of the points in Exercise 17.

19. The polar coordinates of the point $(\sqrt{3}, 1)$ are _____, of the point $(-\sqrt{3}, -1)$ are _____.

20. Convert the rectangular equation $x^2 + y^2 + 6y = 0$ into an equation in polar coordinates.

21. Convert the polar equation $r(1 - \cos\theta) = 1$ into an equation in rectangular coordinates.

22. Sketch the graph of $r = 3 \sin 2\theta$.

23. Find the polar form for the complex number $\frac{1}{2}(\sqrt{3} - i)$.

24. The rectangular form of $2 \operatorname{cis}(\pi/3)$ is _____.

25. Find the product of $3 \operatorname{cis}(\pi/3)$ and $4 \operatorname{cis}(\pi/4)$.

26. If $z_1 = 4 \operatorname{cis}(\pi/4)$ and $z_2 = 2 \operatorname{cis}(\pi/6)$ then $z_1/z_2 =$ _____.

27. State de Moivre's formula.

28. Compute $(2 - 2i)^4$, i^{100}, and $(1 + i)^{24}$.

29. If $z = r \operatorname{cis}\theta$ $(r \neq 0)$, then $z^{1/n} =$ _____.

30. Find the cube roots of i.

TEST FIVE

1. If $z=4-6i$ and $w=3+2i$, calculate

   (a) $z+w$ (b) zw (c) $|z|$ (d) $\overline{zw}$

2. Write the complex number $(3-5i)/(2-7i)$ in the form $a+bi$.

3. (a) Find the rectangular coordinates of the point whose polar coordinates are $(-2, 2\pi/3)$.

   (b) Find a set of polar coordinates for the point whose rectangular coordinates are $(3, -3\sqrt{3})$.

4. Let $z=\sqrt{3}-i$ and $w=1+\sqrt{3}i$.

   (a) Write each in polar form.

   (b) Using the polar form find

            (i) zw (ii) z/w

5. Compute each of the following.

   (a) $(\sqrt{3}+i)^7$

   (b) $[3\ \text{cis}(\pi/12)]^4$

6. Find all of the cube roots of $1-i$ and represent them geometrically.

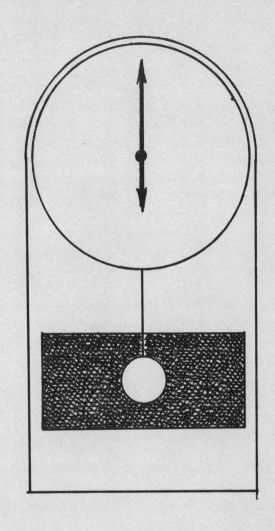

CHAPTER SIX:
VECTORS

6.1 GEOMETRIC REPRESENTATION OF VECTORS

OBJECTIVES

1. *Find the sum (or resultant) of two vectors using geometry.*
2. *Find the scalar multiple of a vector.*

In many applications certain quantities are associated with a specified direction. Such quantities having a *magnitude* and a *direction* are called **vectors**. The notion of vectors evolved out of the study of forces in physics. Today, vectors are used to study many of the basic ideas of algebra, geometry, and trigonometry. The following are some examples of vectors.

A ship travels 100 kilometers to the northwest.

The wind is blowing 20 mph from the east.

An object weighing 250 kilograms is being pulled up a 40° incline.

A vector in the plane can be represented as a directed line segment. We shall use an arrow as shown in Figure 1. The vector is denoted by **AB** where the point A is called the **initial point** and the point B is called the **terminal point**. The length of the directed line segment is called the **magnitude** of the vector **AB** and is denoted by |**AB**|. This is shown in Figure 2.

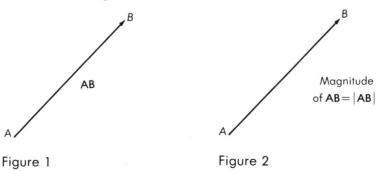

Figure 1   Figure 2

Two vectors **AB** and **CD** are considered **equal**, and we write **AB=CD**, if and only if, they have the same magnitude and direction as shown in Figure 3. A vector may also be denoted by a single boldface letter such as **v**.

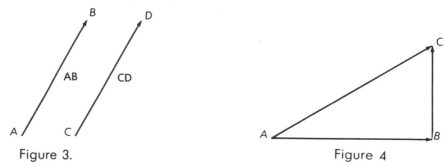

Figure 3.   Figure 4

Suppose a particle moves 4 units east from A to B and then 3 units north from B to C (see Figure 4). The particle then will be 5 units from A, the starting point, in the direction shown. The vector **AC** is called the **sum** of the vectors **AB** and **BC**, and we write

$$AC=AB+BC$$

The sum is also called the **resultant** of the two vectors. To add two vectors, we find a single vector that would have the same effect as the two vectors combined.

In general, if we have two vectors **u** and **v** we can add them in one of two ways.

1. Place the tail of one arrow at the head of the other. Then find the vector that forms the third side of the triangle as shown in Figure 5.

Figure 5

2. Place the tails of the arrows together and construct a parallelogram. The diagonal of the parallelogram starting at the tails of the arrows is the resultant vector as shown in Figure 6.

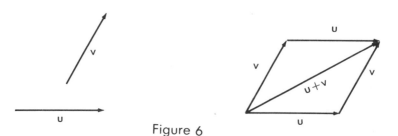

Figure 6

If c is a real number and v is a vector, then the product cv, called a **scalar multiple** of v, is a vector whose magnitude is $|c|$ times the magnitude of v and whose direction is the same as v if $c>0$ and opposite that of v if $c<0$. For example, 2v has the same direction as v with a magnitude twice the magnitude of v, whereas $-3v$ has a direction opposite to that of v with a magnitude three times the magnitude of v (see Figure 7).

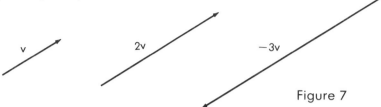

v   2v $-3v$

Figure 7

EXAMPLE 1. A vector **AB** with magnitude 5 is directed due west. A second vector **AD** with a magnitude of 4 has a direction of N30°W. Find the magnitude and direction of the resultant vector.

Solution. Construct the parallelogram with vectors **AB** and **AD** as shown in Figure 8. Clearly, the angle ABC has measure 120° and the length of the line segment BC is 4. Using the Cosine Formula we find

$$|AC|^2 = 5^2 + 4^2 - 2(5)(4)\cos 120°$$
$$= 41 - 20(-\tfrac{1}{2})$$
$$= 41 + 10$$

Therefore, $= 51$
$$|AC| = \sqrt{51}$$

Figure 8

To find the direction of **AC** we use the Sine Formula. Let θ denote the angle CAB. Then

$$(\sin \theta)/4 = (\sin 120°)/\sqrt{51}$$

or

$$\sin \theta = (4/\sqrt{51})\sin 120°$$
$$= (4/\sqrt{51})\sqrt{3}/2$$
$$= 2/\sqrt{17} \approx 0.4851$$

Using Table 1 and interpolation we find that $\theta \approx 29°1'$. Therefore, the direction of AC is N60°59′W.

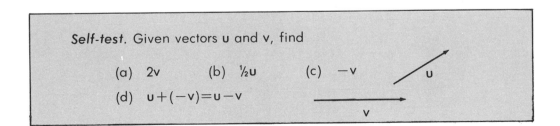

Self-test. Given vectors u and v, find

(a) 2v   (b) ½u   (c) $-v$

(d) $u+(-v)=u-v$

EXERCISES 6.1

In Exercises 1-10, let $A,B,$ and C be three noncollinear points in the plane. Find and illustrate geometrically each of the given vectors.

1. $AB + BC$ 2. $AB - BC$ 3. $2AB + 3BC$

4. $3AB - 3BC$ 5. $-2AB - 2BC$ 6. $-2AC - 2CB$

7. $AB + BC + AC$ 8. $2AB + 3BC - 2AC$ 9. $\frac{1}{2}AC - 3BC + 2CB$

10. $2AB - 3BA + AC$

11. Let u and v be two vectors at rights angles to each other. If $|u| = 3$ and $|v| = 4$, find the resultant of u and v.

In Exercises 12-15, let u and v be the vectors described in Exercise 11. Draw a diagram to illustrate each given vector.

12. $2u + 3v$ 13. $2u + \frac{1}{2}v$ 14. $\frac{1}{3}u - \frac{1}{2}v$ 15. $-2u - 3v$

16. Given two vectors u and v. If $|u| = |v|$, does it follow that $u = v$?

17. Use vectors to prove that the length of the line segment joining the midpoints of two sides of a triangle is one-half the length of the third side and parallel to it.

6.2 ANALYTIC REPRESENTATION OF VECTORS

OBJECTIVES

1. *Express vectors as ordered pairs of real numbers in a Cartesian plane.*
2. *Perform vector addition and scalar multiplication by using components.*
3. *Express a vector in the form $ai+bj$.*
4. *Specify a vector in polar form.*

Suppose a given vector **v** is in a plane with a Cartesian coordinate system. Since the position of **v** may be changed, provided its direction and magnitude are not altered, we place the initial point of the vector at the origin. The terminal point of **v** may then be assigned a coordinate (a,b) as shown in Figure 1. We note that there is a one-to-one correspondence between vectors and ordered pairs of numbers. This permits us to consider a vector in a Cartesian plane as an ordered pair of real numbers instead of directed line segments as was done in Section 6.1.

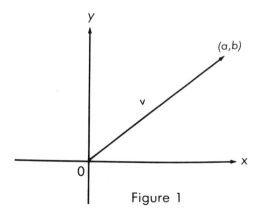

Figure 1

We shall use the symbol $<a,b>$ for the ordered pair representing a vector v. The number a is called the x **component** of v and the number b is called the y **component** of v. The magnitude of $<a,b>$ is, by definition, the distance from the origin to the point (a,b). Thus, if $v=<a,b>$, then

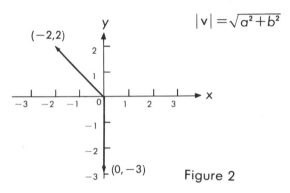

$$|v|=\sqrt{a^2+b^2}$$

Figure 2

EXAMPLE 1. Sketch each of the following vectors. Find the magnitude and the smallest positive angle θ that the vector makes with the positive x-axis.

(a) $u=<-2,2>$ (b) $v=<0,-3>$

Solution. The vectors are sketched in Figure 2. We find

(a) $|u| = \sqrt{(-2)^2+2^2} = 2\sqrt{2};\quad \theta=\arctan(-1)=3\pi/4$

(b) $|v| = \sqrt{0^2+3^2} =3;\qquad\qquad \theta=3\pi/2$

If $u=<a,b>$ and $v=<c,d>$, then it can be shown that

$$u+v=<a,b>+<c,d>=<a+c,\ b+d>$$

and

$$k<a,b>=<ka,\ kb>$$

where k is a scalar (see Figures 3 and 4).

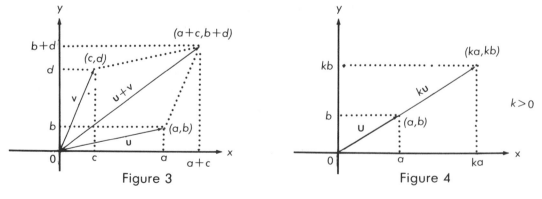

Figure 3   Figure 4

EXAMPLE 2. If $u=<-3,5>$ and $v=<6,-4>$, find each of the following:

(a) $u+v$ (b) $-2u$ (c) $2u+3v$

Solution.

(a) $u+v = <-3,5>+<6,-4>$
$= <-3+6, 5-4>=<3,1>$

(b) $-2u = -2<-3,5>=<6,-10>$

(c) $2u+3v = 2<-3,5>+3<6,-4>$
$= <-6,10>+<18,-12>$
$= <-6+18, 10-12>=<12,-2>$

By definition, the **zero vector**, denoted by 0, corresponds to $<0,0>$. Also, if $u=<a,b>$, then we define $-u=<-a,-b>$. With these definitions, the following properties are readily established:

$$u+v= v+u$$
$$u+(v+w) = (u+v)+w$$
$$u+0= u$$
$$u+(-u) = 0$$

If **u** and **v** are two vectors, we define subtraction as follows:

$$u-v=u+(-v)$$

Thus, if $u=<a,b>$ and $v=<c,d>$ then

$$u-v=<a,b>+<-c,-d>=<a-c, b-d>$$

Self-test.   Let $u=<4,-2>$ and $v=<5,3>$. Find

(a) $u+v=$ _____   (b) $-3u=$ _____   (c) $u-v=$ _____

Now, let $v=<a,b>$. Then we may write **v** as follows:

$$v=<a,b>=<a,0>+<0,b>$$
$$=a<1,0>+b<0,1>$$

ANSWERS:

(a) $<9,1>$ (b) $<-12,6>$ (c) $<-1,-5>$

Since the magnitude of each vector $<1,0>$ and $<0,1>$ is one unit, they are called **unit vectors**. The following symbols are used for these vectors:

$$i=<1,0>, \ j=<0,1>$$

Thus, we may write v as:

$$v=ai+bj$$

We call a the **horizontal component** and b the **vertical component** of the vector v.

EXAMPLE 3. If $u=2i+j$ and $v=3i-2j$, find $3u-4v$.

Solution.

$$\begin{aligned} 3u-4v &= 3(2i+j)-4(3i-2j) \\ &= 6i+3j-12i+8j \\ &= -6i+11j \end{aligned}$$

Self-test. If $v=i-2j$ and $u=3i+j$, find each of the following.

(a) $|v| =$ _____   (b) $2u-3v=$ _____

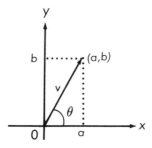

y

b .... (a,b)

v

θ

0   a

x

Figure 5

A vector may be specified in **polar form** by giving its magnitude and its direction. Consider the vector $v=ai+bj$ (see Figure 5). The vector v can be expressed in terms of $|v|$ and θ as follows:

$$a=|v| \cos \theta, \quad b=|v| \sin \theta$$

Thus

$$v=|v| \cos \theta i+|v| \sin \theta \ j=|v|(\cos \theta i-\sin \theta j)$$

The polar form of the vector v is then given by:

$$<|v|, \ \arctan b/a>$$

ANSWERS:

EXAMPLE 4. Find the polar form for the vector $v=<-2,3>$.

Solution. First we find $|v|=\sqrt{(-2)^2+3^2}=\sqrt{13}$ and from Figure 6, we see the reference angle θ.

$$\theta=\arctan(-\tfrac{3}{2})\approx123°40'$$

Thus, the polar form for the vector is

$$<\sqrt{13},\ 123°40'>$$

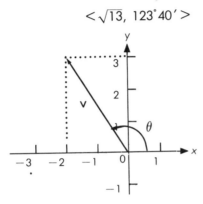

Figure 6

EXERCISES 6.2

In Exercises 1-6, find the magnitude and the smallest positive angle that the vector makes with the positive x-axis.

1. $u=<3,3>$ 2. $v=<2,0>$ 3. $w=<-4,4>$

4. $u=<-1,-1>$ 5. $v=<0,5>$ 6. $w=<2,-2>$

In Exercises 7-14, find $u+v$ and $u-v$ and sketch the vectors corresponding to u, v, $u+v$ and $u-v$.

7. $u=<2,3>$, $v=<1,-4>$ 8. $u=<3,5>$, $v=<-1,5>$

9. $u=<0,-4>$, $v=<6,5>$ 10. $u=<1,2>$, $v=<3,5>$

11. $u=<-2,-1>$, $v=<-3,-4>$ 12. $u=<3,0>$, $v=<0,3>$

13. $u=<5,1>$, $v=<1,5>$ 14. $u=<-2,3>$, $v=<0,0>$

In Exercises 15-30, let $u=<2,-3>$, $v=<3,5>$, and $w=<-5,6>$. Find each of the given expressions.

15. $u+(v+w)$ 16. $(u+v)+w$ 17. $2u+2v$

18. $2(u+v)$ 19. $-3v+3w$ 20. $-3(v-w)$

21. $3u+2v$ 22. $-2u-5w$ 23. $u+(2v-3w)$

24. $2u-(v+w)$ 25. $|u|+|v|$ 26. $|u|+|w|$

27. $|3u-v|$ 28. $|3u|-|v|$ 29. $|2v+3w|$

30. $|2v|+|3w|$

31.   Prove that for two vectors u and v, $u+v=v+u$.

32.   Prove that $u+0=u$.

33.   Prove that $u+(-u)=0$.

In Exercises 34-37, find polar form for each vector.

34. $v=<3,4>$ 35. $v=4i+3j$ 36. $u=10i-15j$ 37. $u=-15i+10j$

In Exercises 38-40, find rectangular form for each vector.

38. $<5,30°>$ 39. $<4,60°>$ 40. $<3,135°>$

6.3 INNER PRODUCT AND VECTOR APPLICATIONS

OBJECTIVES

1. *Given two vectors define the inner product.*
2. *Solve applied problems involving vectors.*

We now consider the problem of finding the angle between two vectors $u=<a,b>$ and $v=<c,d>$. Let θ be the angle whose sides are u and v, with $0\leq\theta\leq180°$. We define an operation on vectors, the **inner product** (also called the **dot product**), denoted by $u\bullet v$ as follows:

$$u\bullet v=|u||v|\cos\theta \qquad (1)$$

where θ is the angle between u and v (see Figure 1). The inner product combines two vectors to produce a scalar quantity. If either $u=0$ or $v=0$, the angle between u and v is not defined and in this case, we have by definition

$$u\bullet v=0$$

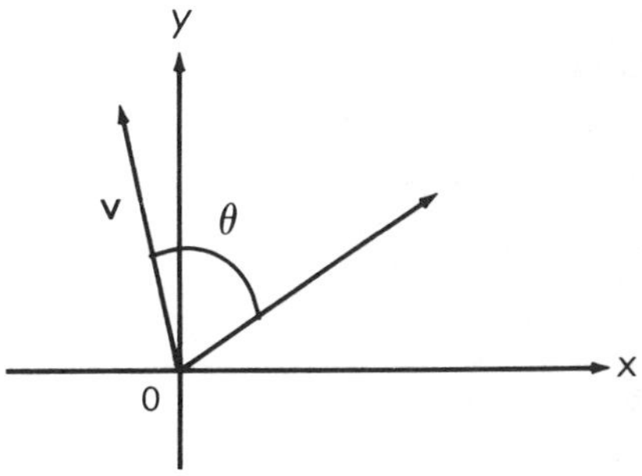

Figure 1

Using the definition of inner product it can be easily shown that

$$\mathbf{u} \cdot \mathbf{v} = \mathbf{v} \cdot \mathbf{u} \qquad\qquad (2)$$

$$\mathbf{v} \cdot \mathbf{v} = |\mathbf{v}|^2 \qquad\qquad (3)$$

$$\mathbf{u} \cdot \mathbf{v} = 0, \text{ if and only if } \mathbf{u} \text{ and } \mathbf{v} \text{ are perpendicular} \qquad (4)$$

EXAMPLE 1. Let $\mathbf{u}$ and $\mathbf{v}$ be two vectors such that $|\mathbf{u}| = 2$, $|\mathbf{v}| = 4$ and the angle between them is 60°. Find $\mathbf{u} \cdot \mathbf{v}$.

Solution.
$$\mathbf{u} \cdot \mathbf{v} = |\mathbf{u}||\mathbf{v}| \cos \theta = (2)(4) \cos 60° = 8(\tfrac{1}{2}) = 4$$

Self-test. Let $|\mathbf{u}| = 5$ and $|\mathbf{v}| = 8$. If θ is the angle between $\mathbf{u}$ and $\mathbf{v}$, then

(a)   For $\theta = 45°$, $\mathbf{u} \cdot \mathbf{v} =$ _____

(b)   For $\theta = 90°$, $\mathbf{u} \cdot \mathbf{v} =$ _____

(c)   For $\theta = 180°$, $\mathbf{u} \cdot \mathbf{v} =$ _____

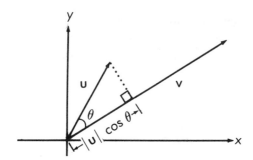

Figure 2

Consider the vectors $\mathbf{u}$ and $\mathbf{v}$ shown in Figure 2. The product

$$|\mathbf{u}| \cos \theta$$

is the length of the projection of $\mathbf{u}$ on $\mathbf{v}$ if $0 \le \theta \le 90°$ and the negative of the length if $90° \le \theta \le 180°$. The quantity $|\mathbf{u}| \cos \theta$ is called the **scalar projection** of $\mathbf{u}$ on $\mathbf{v}$.

EXAMPLE 2. Find the scalar projection of $\mathbf{u}$ on $\mathbf{v}$ of the vectors $\mathbf{u}$ and $\mathbf{v}$ in Example 1.

Solution. $|\mathbf{u}| \cos \theta = (\mathbf{u} \cdot \mathbf{v})/|\mathbf{v}| = \tfrac{4}{4} = 1.$

ANSWERS:

Self-test. Find the scalar projection of **u** on **v** of the vectors given in the last Self-test.

(a) $|\mathbf{u}|\cos\theta=$ _____

(b) $|\mathbf{u}|\cos\theta=$ _____

(c) $|\mathbf{u}|\cos\theta=$ _____

The inner product may also be expressed in terms of the components of the vectors involved. Let

$$\mathbf{u}=a_1\mathbf{i}+a_2\mathbf{j} \quad \text{and} \quad \mathbf{v}=b_1\mathbf{i}+b_2\mathbf{j}$$

and suppose θ is the angle between **u** and **v** as shown in Figure 3. Using the Cosine Formula, we have

$$|\mathbf{v}-\mathbf{u}|^2=|\mathbf{u}|^2+|\mathbf{v}|^2-2|\mathbf{u}||\mathbf{v}|\cos\theta$$

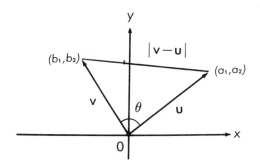

Figure 3

so that

$$
\begin{aligned}
|\mathbf{u}||\mathbf{v}|\cos\theta &= \tfrac{1}{2}(|\mathbf{u}|^2+|\mathbf{v}|^2-|\mathbf{v}-\mathbf{u}|^2)\\
&= \tfrac{1}{2}[a_1^2+a_2^2 + b_1^2+b_2^2-(b_1-a_1)^2-(b_2-a_2)^2]\\
&= \tfrac{1}{2}[2(a_1b_1+a_2b_2)]\\
&= a_1b_1+a_2b_2
\end{aligned}
\tag{5}
$$

By the definition of the inner product

$$\mathbf{u}\cdot\mathbf{v}=|\mathbf{u}||\mathbf{v}|\cos\theta \tag{6}$$

Thus, from Equations 5 and 6 we get

$$\mathbf{u}\cdot\mathbf{v}=a_1b_1+a_2b_2 \tag{7}$$

ANSWERS:

(a) $\sqrt{2}/2$ (b) 0   (c) -5

Also, from Equations 6 and 7 we find

$$\cos \theta = \frac{a_1b_1 + a_2b_2}{\sqrt{a_1^2 + a_2^2} \ \sqrt{b_1^2 + b_2^2}} \tag{8}$$

Using Equation 7 we can readily show that

$$u \cdot (v + w) = u \cdot v + u \cdot w \tag{9}$$

EXAMPLE 3. Compute each of the following inner products.

   (a) $i \cdot i$ (b) $i \cdot j$ (c) $<2,1> \cdot <-3,8>$

Solution.

 (a) $i \cdot i = (1i + 0j) \cdot (1i + 0j) = (1)(1) + (0)(0) = 1$
 (b) $i \cdot j = (1i + 0j) \cdot (0i + 1j) = (1)(0) + (0)(1) = 0$
 (c) $<2,1> \cdot <-3,8> = (2i + j) \cdot (-3i + 8j) = (2)(-3) + (1)(8) = 2$

EXAMPLE 4. Find the angle θ between the vectors $u = 3i + 4j$ and $v = 5i + j$.

Solution. First we find

$$|u| = \sqrt{3^2 + 4^2} = 5$$
$$|v| = \sqrt{5^2 + 1^2} = \sqrt{26}$$
$$u \cdot v = (3)(5) + (4)(1) = 19$$

Using Equation 8, we get

$$\cos \theta = (u \cdot v)/(|u||v|) = 19/(5\sqrt{26})$$

Thus,

$$\theta = \arccos[(19\sqrt{26})/130]$$

Self-test.

 (a) Compute $j \cdot j =$ _____

 (b) Compute $<-1,2> \cdot <3,-4> =$ _____

 (c) Find the angle θ between $u = <1,-2>$ and j: $\theta =$ _____

ANSWERS:

The next three examples illustrate the applications of vectors to solving problems in such areas as geometry and trigonometry.

EXAMPLE 5. Show that the diagonals of a rhombus are perpendicular to each other.

Solution. In Figure 4, ABCD is a rhombus. Let **AB**=**u** and **AD**=**v**. Then the diagonals may be represented by **AC**=**u**+**v**, and **DB**=**u**−**v**. Now, **AC** and **DB** are perpendicular if and only if

$$\textbf{AC}\cdot\textbf{DB}=0$$

That is, we need to show that

$$(\textbf{u}+\textbf{v})\cdot(\textbf{u}-\textbf{v})=0$$

First we note that for the rhombus $|\textbf{u}|=|\textbf{v}|$. Hence, $|\textbf{u}|^2=|\textbf{v}|^2$ or $|\textbf{u}|^2-|\textbf{v}|^2=0$. Using Equations 2,3 and 9 we find

$$\begin{aligned}(\textbf{u}+\textbf{v})\cdot(\textbf{u}-\textbf{v}) &= (\textbf{u}+\textbf{v})\cdot\textbf{u}-(\textbf{u}+\textbf{v})\cdot\textbf{v}\\ &= \textbf{u}\cdot\textbf{u}+\textbf{u}\cdot\textbf{v}-\textbf{u}\cdot\textbf{v}-\textbf{v}\cdot\textbf{v}\\ &= \textbf{u}\cdot\textbf{u}-\textbf{v}\cdot\textbf{v}\\ &= |\textbf{u}|^2-|\textbf{v}|^2\\ &= 0\end{aligned}$$

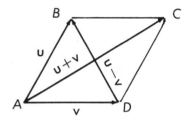

Figure 4

Therefore, the diagonals of the rhombus are perpendicular.

EXAMPLE 6. Use the inner product to prove the identity

$$\cos(\alpha-\beta)=\cos\alpha\cos\beta+\sin\alpha\sin\beta$$

Solution. In Figure 5 we illustrate the vectors involved relative to the angles α, β and $\alpha-\beta$. Let **u** and **v** be unit vectors. Then

$$\textbf{u}=\cos\alpha\textbf{i}+\sin\alpha\textbf{j} \text{ and } \textbf{v}=\cos\beta\textbf{i}+\sin\beta\textbf{j}$$

with

$$|\textbf{u}|=|\textbf{v}|=1$$

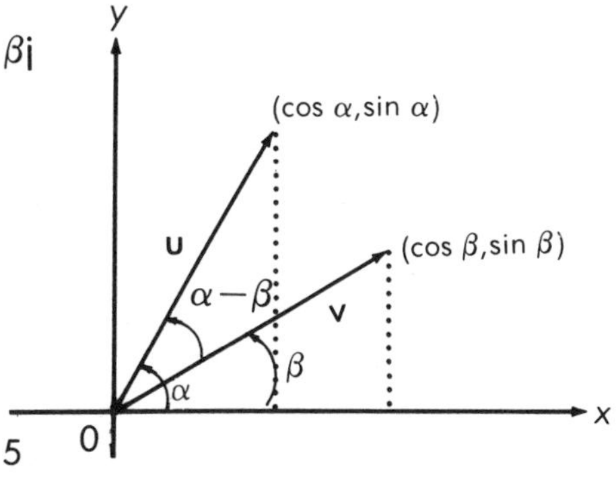

Figure 5

By the definition of the inner product

$$\mathbf{u} \cdot \mathbf{v} = |\mathbf{u}||\mathbf{v}| \cos(\alpha - \beta) = \cos(\alpha - \beta)$$

From Equation 7,

$$\mathbf{u} \cdot \mathbf{v} = \cos \alpha \cos \beta + \sin \alpha \sin \beta$$

Hence,

$$\cos(\alpha - \beta) = \cos \alpha \cos \beta + \sin \alpha \sin \beta$$

EXAMPLE 7. A boat heads in the direction of N20°W propelled by a force of 500 pounds. A current from S40°W exerts a force of 200 pounds on the boat. Find the resultant force and the direction in which the boat is moving.

Solution. First we make a drawing as in Figure 6. The force of the boat is represented by **OA**, the force of the current is represented by **OB**, and the resultant force is represented by **OC**. Using the Cosine Formula, in triangle OCB, we have

$$|\mathbf{OC}|^2 = |\mathbf{OB}|^2 + |\mathbf{BC}|^2 - 2|\mathbf{OB}||\mathbf{BC}| \cos 120°$$

$$= (200)^2 + (500)^2 - 2(200)(500)(-\tfrac{1}{2})$$

$$= 280,000$$

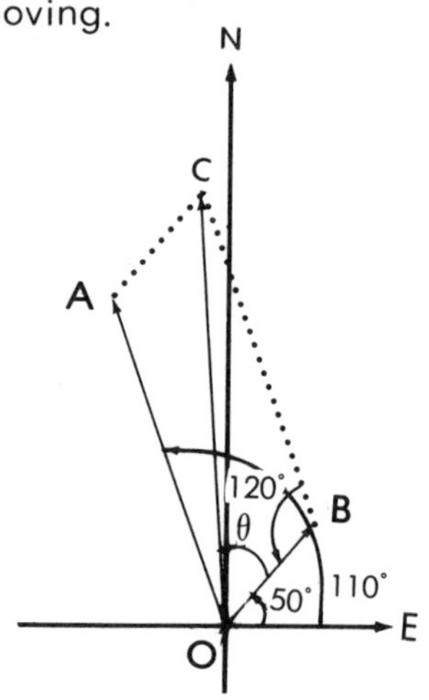

Figure 6

Thus the magnitude of the resultant force is

$$|\mathbf{OC}| = \sqrt{280,000} \approx 530 \text{ pounds}$$

To find the angle between **OC** and **OB** we use the Sine Formula in triangle OCB.

$$500/\sin \theta = 530/\sin 120°$$

or

$$\sin \theta = (500 \sin 120°)/530 = (500/530)(\sqrt{3}/2)$$

Thus

$$\theta = 54°47'$$

Therefore, the resultant **OC** has a direction of N14°47'W.

EXERCISES 6.3

1. Find $\mathbf{u} \cdot \mathbf{v}$ given that $|\mathbf{u}| = 6$ and $|\mathbf{v}| = 7$ and the angle between $\mathbf{u}$ and $\mathbf{v}$ is:

    (a) $\theta = 30°$ (b) $\theta = 60°$ (c) $\theta = 135°$ (d) $\theta = 150°$

2. Find the scalar projection of $\mathbf{u}$ on $\mathbf{v}$ for each pair of vectors in Exercise 1.

3. Prove that $\mathbf{v} \cdot \mathbf{v} = |\mathbf{v}|^2$.

4. Prove that $\mathbf{u} \cdot \mathbf{v} = 0$ if and only if $\mathbf{u}$ and $\mathbf{v}$ are perpendicular.

5. Prove that $\mathbf{u} \cdot (\mathbf{v} + \mathbf{w}) = \mathbf{u} \cdot \mathbf{v} + \mathbf{u} \cdot \mathbf{w}$.

In Exercises 6-13, compute the inner products.

6. $<0,2> \cdot <1,3>$ 7. $<1,0> \cdot <1,3>$

8. $<1,1> \cdot <-1,2>$ 9. $<-1,-1> \cdot <1,1>$

10. $<3,4> \cdot <-4,3>$ 11. $<1,2> \cdot <2,1>$

12. $<-2,3> \cdot <1,-3>$ 13. $<1,-3> \cdot <-3,2>$

In Exercises 14-20, compute the angle between the vectors $\mathbf{u}$ and $\mathbf{v}$.

14. $\mathbf{u} = 3\mathbf{i}, \quad \mathbf{v} = 2\mathbf{j}$ 15. $\mathbf{u} = -2\mathbf{i}, \quad \mathbf{v} = 3\mathbf{j}$

16. $\mathbf{u} = 5\mathbf{i}, \quad \mathbf{v} = \mathbf{i} - \mathbf{j}$ 17. $\mathbf{u} = \mathbf{i} - 2\mathbf{j}, \quad \mathbf{v} = -2\mathbf{i} + \mathbf{j}$

18. $\mathbf{u} = \mathbf{i} + \mathbf{j}, \quad \mathbf{v} = -\mathbf{i} + 7\mathbf{j}$ 19. $\mathbf{u} = \mathbf{i} + 3\mathbf{j}, \quad \mathbf{v} = 3\mathbf{i} + \mathbf{j}$

20. $\mathbf{u} = 3\mathbf{i} + 4\mathbf{j}, \quad \mathbf{v} = -5\mathbf{i} - 2\mathbf{j}$

21. If $\mathbf{u} = <a,b>$ and $\mathbf{v} = <c,d>$, show that
$$|\mathbf{u} - \mathbf{v}|^2 = (a-c)^2 + (b-d)^2$$

22. Prove that $|\mathbf{u} + \mathbf{v}|^2 = |\mathbf{u}|^2 + |\mathbf{v}|^2 + 2\mathbf{u} \cdot \mathbf{v}$.

23. Prove that $|\mathbf{u} + \mathbf{v}|^2 + |\mathbf{u} - \mathbf{v}|^2 = 2(|\mathbf{u}|^2 + |\mathbf{v}|^2)$.

24. Prove that $|\mathbf{u} + \mathbf{v}|^2 - |\mathbf{u} - \mathbf{v}|^2 = 4\mathbf{u} \cdot \mathbf{v}$.

25. Use vectors to prove that the diagonals of a rectangle are equal.

26. Use vectors to prove that the altitudes of a triangle are concurrent.

27. Use vectors to prove that the median to the base of an isosceles triangle is perpendicular to the base.

28. Two forces of 24 pounds and 10 pounds act on an object. If the two forces are at right angles, find the magnitude of the resultant and the angle it makes with the larger force.

29. Two forces of 210 pounds and 150 pounds act on an object. If the angle between the two forces is 50°, find the resultant, giving the angle it makes with the smaller force.

30. An airplane heads in a direction of N10°W at 200 mph airspeed. If a wind is blowing 40 mph from S50°W, find the speed of the airplane over the ground and its direction.

REVIEW OF CHAPTER SIX

1. Given two vectors u and v, indicate how you would find their sum (or resultant).

2. A vector **AB** with magnitude 3 is directed due east. A second vector **CD** with magnitude 5 is directed due north. Find the magnitude and direction of the resultant vector.

3. For the vector $v = <a,b>$, a is called _____ and b is called _____.

4. The magnitdue of the vector $u = <c,d>$ is _____.

5. Find the magnitude and the smallest positive angle that the vector $v = <3,-3>$ makes with the positive x-axis.

6. If $u = <a,b>$ and $v = <c,d>$ then

    (a) $ku =$ _____

    (b) $u+v =$ _____

    (c) $u+(-u) =$ _____

    (d) $u+0 =$ _____

7. Let $u=<-1,2>$ and $v=<3,-5>$. Find

    (a) $u+v=$ _____
    (b) $-2u=$ _____
    (c) $u-v=$ _____
    (d) $2u-3v=$ _____
    (e) $|u|=$ _____
    (f) $|u+v|=$ _____

8. If $u=<a,b>$, write u in the form using the unit vectors i and j. $u=$____.

9. If $u=3i-j$ and $v=2i+5j$, find

    (a) $|u|=$ _____
    (b) $u+v=$ _____
    (c) $|u|+|v|=$ _____
    (d) $|u+v|=$ _____
    (e) $u-v=$ _____
    (f) $2u-5v=$ _____

10. Find polar form for the vector $u=<-2,2>$.

11. Find rectangular form for the vector $v=<2,150°>$.

12. Define inner product.

13. Is it always true that $u•v=v•u$?

14. Complete the sentence: $u•v=0$ if and only if u and v are _____
    provided u and v are not the zero vectors.

15. Let $|u|=5$ and $|v|=6$. If the angle between them is $135°$ find $u•v$.

16. If $u=<2,1>$ and $v=<-1,3>$ find $u•v$.

17. Find the scalar projection of u on v if $|u|=3$ and $|v|=5$ and the angle
    between u and v is $60°$.

18. Show that $(v+w)•u=u•v+u•w$.

19. Find the angle between the vectors $u=i+j$ and $v=2i-2j$.

TEST SIX

1. Let A, B, and C be three noncollinear points in the plane. Find

   (a) **AB+BC**   (b) **AB+CB**   (c) **−AC−CB**   (d) **2AB−3CB**

2. Determine the components of the vector v if $|v|=5$ and the angle between v and the positive x-axis is 120°.

3. Let $u=<-4,3>$ and $v=<4,-2>$. Determine each of the following:

   (a) $|u|=$ _____

   (b) $u+v=$ _____

   (c) $u-v=$ _____

   (d) $3u-2v=$ _____

   (e) $|u+v|=$ _____

   (f) $|u|-|v|=$ _____

   (g) $|u|^2=$ _____

4. Let $u=<-5,3>$ and $v=<2,4>$. Find:

   (a) $u \cdot v=$ _____

   (b) $2u \cdot 3v=$ _____

   (c) $u \cdot u=$ _____

   (d) $(u-v) \cdot (u+v)=$ _____

5. Find the angle between u and v in Problem 4.

6. Let $u=4i+3j$ and $v=2ci-5j$. Find the number c so that u and v are perpendicular to each other.

7. Find a unit vector u that is perpendicular to the vector $v=3i-4j$.

8. A vector of magnitude 20 is inclined 60° with the horizontal. Find its horizontal and vertical components.

9. Two forces of magnitude 30 pounds and 40 pounds are acting on an object. If the respective directions are N70°W and S10°E, find the magnitude and the direction of the resultant.

10. Use vectors to derive the distance formula between the points (x_1,y_1) and (x_2,y_2).

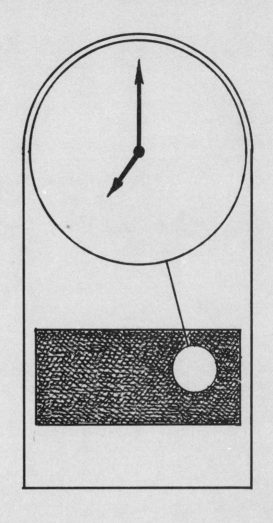

CHAPTER SEVEN:
CONIC SECTIONS

7.1 INTRODUCTION

OBJECTIVES

1. Give a brief historical background for conic sections.
2. Discuss the geometry of the conic sections.

Galileo proved that under ideal conditions the path of a projectile is a mathematical curve called a **parabola**. In the construction of the Golden Gate Bridge in San Francisco, cables are hung between two towers to suspend the bridge. The cables form a curve that is very close to the shape of a parabola. Certain properties of the parabola make it useful in the design of such items as equipment for radio and television, automotive headlights, and parabolic mirrors.

The famous Polish astronomer, Nicholas Copernicus (1743-1543), discarded the old Ptolemaic theory and claimed that the sun, rather than the earth, was the center of our solar system. He used the **circle** as the basic curve to describe the motion of the planets. A century later, Johannes Kepler (1571-1630), introduced the famous laws of planetary motion. In his first law, Kepler stated that the path of each planet was an **ellipse** with the sun at one focus. We note that the orbits of man-made satellites are also ellipses.

Another plane curve that has many applications in today's world is the **hyperbola**. When an airplane flies at a speed greater than the speed of sound (about 350 meters per second) a shock wave is created which is heard as a sonic boom. The shock wave is in the shape of a cone. The shape formed as the cone hits the ground is known as a hyperbola. Properties of the hyperbola make it useful in the battlefield to locate an enemy gun position. At sea, a ship receiving signals from two unknown stations on shore can plot its position as the intersection of the hyperbolas formed by the signals.

The early Greek mathematicians were familiar with the circle, the ellipse, the parabola, and the hyperbola. These plane curves, called **conic sections**, were discovered by the Platonic school. Euclid's book, The *Conics*, formed the basis of the first half of the work of Apollonius who lived in the third century B.C. Apollonius came from Perga in Asia Minor and studied at Alexandria in Eygpt. He introduced the names parabola, ellipse, and hyperbola and showed that these curves could be produced by slicing a cone in different directions; hence the name conic sections (or simply conics).

The conic sections were introduced into the realm of algebra by Descartes, Fermat, and Wallis. They defined the conics as curves of second degree equations in x and y of the form

$$ax^2 + bxy + cy^2 + dx + ey + f = 0$$

where a, b, c, d, e, and f are constants.

We shall now discuss the geometry of the conic sections and introduce some basic terminology involved in describing these plane curves analytically. Consider a cone having two nappes extending indefinitely in both directions. In Figure 1 we show a portion of such a cone. The upper and lower nappes meet at the **vertex** of the cone. Any line in the cone containing the vertex is called an **element** (or **generator**) of the cone.

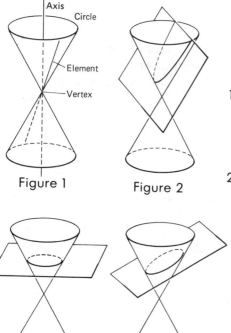

Figure 1   Figure 2

Figure 3

We have the following cases.

1. If the cutting plane is parallel to one and only one of the elements of the cone, a **parabola** is obtained. (see Figure 2).

2. If a cutting plane is parallel to none of the elements of the cone, an **ellipse** is obtained. In this case, the cutting plane intersects each of the elements (see Figure 3). A special case of the ellipse is the **circle**, which is formed if the cutting plane intersects each element and is perpendicular to the axis of the cone.

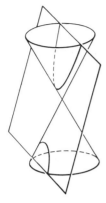

Figure 4

3. If the cutting plane is parallel to two elements of the cone, a **hyperbola** is obtained (see Figure 4).

Degenerate cases of the conic sections are obtained as follows:

i.   If the cutting plane contains the vertex and one element of the cone, a straight line is obtained. This is known as a degenerate parabola.

ii. If the cutting plane contains the vertex but does not contain an element of the cone, a point is obtained. This is known as a degenerate ellipse.

iii. If the cutting plane contains the vertex and two elements of the cone, two intersecting lines are obtained. This is known as a degenerate hyperbola.

Now we can discuss the conic sections analytically as curves in the plane. The following definition gives a property that is shared by the conic sections.

> **DEFINITION 7.1**   A *conic section* (or simply a *conic*) is the set of all points P in a plane such that the distance of P from a fixed point is in a constant ratio to the distance of P from a fixed line which does not contain the fixed point.

The fixed point is called the **focus** and the fixed line is called the **directrix**. The constant ratio is called the **eccentricity** of the conic and is denoted by e.

For the nondegenerate conic the eccentricity e is nonnegative, since it is the ratio of two distances. It turns out that

i.   for the ellipse $0 \leq e < 1$
ii. for the parabola $e = 1$
iii. for the hyperbola $e > 1$

When $e = 0$ we have a circle.

We shall see that the parabola has one focus and one directrix while the ellipse and the hyperbola have two **foci** (plural of focus) and two directrices. The **principal axis** of a conic is the line passing through the focus and perpendicular to its directrix. The points of intersection of the principal axis and the conic are called the **vertices** of the conic. The parabola has one vertex, while the ellipse and hyperbola each have two vertices. For the ellipse and hyperbola, the point that lies halfway between the two vertices is called the **center** of the conic. Thus, the parabola has no center.

7.2 THE CIRCLE

OBJECTIVES

1. Define a circle and derive an equation describing it.
2. Find an equation of a circle, given its center and radius.
3. Given an equation of a circle, find its center and the radius.
4. Find an equation of a circle with a given center and tangent to a given line.

In this section, we shall consider one of the simplest mathematical curves, the circle. We define a circle as follows.

> **DEFINITION 7.2** The *circle* is the set of all points in the plane that are at the same distance from a fixed point in the plane. The fixed point is called the *center* of the circle and the fixed distance is called the *radius*.

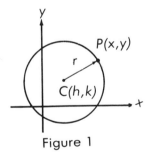

Figure 1

Let us consider a circle with center at the point $C(h,k)$ and radius r (see Figure 1). Then the point $P(x,y)$ belongs to the circle if and only if

$$d(P,C)=r$$

or, using the distance formula,

$$\sqrt{(x-h)^2+(y-k)^2}=r \tag{1}$$

Squaring both sides of Equation 1, we obtain

$$(x-h)^2+(y-k)^2=r^2 \tag{2}$$

Equation 2 is called the equation of the circle with center (h,k) and radius r. From Equation 2, we get

$$x^2-2hx+h^2+y^2-2ky+k^2-r^2=0$$

or

$$x^2+y^2-2hx-2ky+h^2+k^2-r^2=0$$

or

$$x^2+y^2+ax+by+c=0 \tag{3}$$

where $a=-2h$, $b=-2k$ and $c=h^2+k^2-r^2$.

EXAMPLE 1. Find the equation of a circle with center at $(3,-4)$ and radius 6.

Solution. Here, $h=3$, $k=-4$, and $r=6$. Substituting these into Equation 2, we obtain

$$(x-3)^2+[y-(-4)]^2=6^2$$

or

$$(x-3)^2+(y+4)^2=36$$

Squaring and combining terms we get

$$x^2+y^2-6x+8y-11=0$$

The circle is shown in Figure 2.

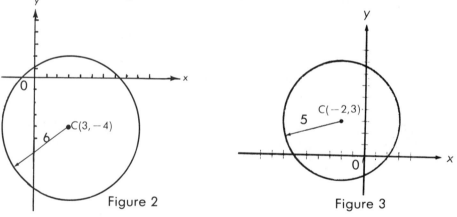

Figure 2   Figure 3

EXAMPLE 2. Find the center and radius of the circle defined by

$$x^2+y^2+4x-6y-12=0 \tag{4}$$

Solution. Equation 4 may be written in the form

$$(x^2+4x)+(y^2-6y)=12$$

Completing the square for the terms involving x and the terms involving y, we get

$$(x^2+4x+4)+(y^2-6y+9)=12+4+9$$

or

$$(x+2)^2+(y-3)^2=25$$

From Equation 2, it is easily seen that we have the equation of a circle with its center at $(-2,3)$ and radius $r=\sqrt{25}=5$. The circle is shown in Figure 3.

Self-test.

(a) An equation of the circle with its center at $(-2,5)$ and radius of 4 is _____.

(b) Find the center and the radius of the circle defined by
$$x^2+y^2-2x+4y=4$$
The center is at _____ and the radius is _____.

EXAMPLE 3. Find the equation of the circle passing through the points $P(-4,1)$, $Q(5,4)$, and $R(1,6)$. Locate the center and determine the radius.

Solution. Let the equation of the circle be
$$x^2+y^2+ax+by+c=0 \tag{5}$$
Then the points P,Q, and R satisfy Equation 5. Substitution yields

$P(-4,1)$:	$16+1-4a+b+c=0$
$Q(5,4)$:	$25+16+5a+4b+c=0$
$R(1,6)$:	$1+36+a+6b+c=0$

These are three equations in the three unknowns a, b, and c. We solve these simultaneously to obtain
$$a=-2,\ b=-2,\ c=-23$$
Substituting the values of a, b, and c into Equation 5, we get
$$x^2+y^2-2x-2y-23=0$$
or
$$(x^2-2x+1)+(y^2-2y+1)=23+1+1$$
or
$$(x-1)^2+(y-1)^2=25$$
Thus, the center is at the point $C(1,1)$ and the radius $r=\sqrt{25}=5$ (see Figure 4).

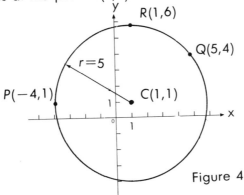

Figure 4

ANSWERS:

(b) $(1,-2),\ r=3$ (a) $(x+2)^2+(y-5)^2=16$

EXAMPLE 4. Find an equation of the circle with its center at $C(-2,1)$ and tangent to the line L defined by $2x+y=1$.

Solution. (See Figure 5). Let r be the radius of the circle. Then r is the distance from C to the line L. Thus we have

$$r= \frac{|(-2)(2)+(1)(1)-1|}{\sqrt{2^2+1^2}} = \frac{4}{\sqrt{5}}$$

and the equation of the circle is

$$(x+2)^2+(y-1)^2=(4/\sqrt{5})^2$$

or

$$x^2+4x+4+y^2-2y+1=\tfrac{16}{5}$$

or

$$5x^2+5y^2+20x-10y+9=0$$

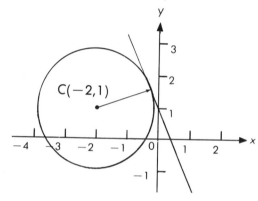

Figure 5

NOTE: It can be shown that the distance from the point (h,k) to the line defined by $ax+by=c$ is given by $|ah+bk-c|/\sqrt{a^2+b^2}$.

EXERCISES 7.2

In Exercises 1-6, find an equation of a circle with its center at C and radius r.

1. $C(-1,2)$, $r=4$ 2. $C(3,-4)$, $r=2$ 3. $C(-2,-4)$, $r=5$

4. $C(1,1)$, $r=6$ 5. $C(-\frac{1}{2},\frac{1}{2})$, $r=\sqrt{5}$ 6. $C(3,-\frac{1}{3})$, $r=\sqrt{10}$

In Exercises 7-10, find the center and radius of the given circle. Sketch the graph.

7. $x^2+y^2-6x=16$ 8. $2x^2+2y^2=8$

9. $x^2+y^2+4x-6y=23$ 10. $x^2+y^2-6x-8y+12=0$

In Exercises 11-22, find an equation of a circle which satisfies the given conditions. Sketch the graph.

11. Center at $(3,-4)$ and containing the point $(-1,5)$.

12. The diameter is a line segment whose endpoints are $(3,1)$ and $(7,5)$.

13. Center at $(-1,3)$ and tangent to the line defined by $y=-1$.

14. Center at $(2,2)$ and tangent to the line defined by $x=5$.

15. Center at $(1,2)$ and tangent to the line defined by $x-y=3$.

16. Center at $(1,2)$ and tangent to the line defined by $3x+y=-2$.

17. Tangent to the line defined by $2x+y=2$ at the point $(1,0)$ and containing the point $(2,4)$.

18. Tangent to the line defined by $2x+y=2$ at the point $(1,0)$, with a radius of 2. Note, there are two possible circles in this case.

19. Tangent to both axes, has a radius of 5 and is in the first quadrant.

20. The center is on the y-axis and passing through the points $(3,4)$ and $(1,2)$.

21. Center on the x-axis and passing through the points $(1,\sqrt{8})$ and $(2,-3)$.

22. Center on the line defined by $x+y=2$ and passing through the points $(1,3)$ and $(-1,1)$.

In Exercises 23-25, find an equation of the circle containing the given three points. Find the center and radius of the circle and sketch the graph.

23. $P(0,1)$, $Q(1,0)$, $R(2,2)$ 24. $P(1,2)$, $Q(-5,2)$, $R(-3,4)$

25. $P(1,-3)$, $Q(2,-2)$, $R(-2,6)$

7.3 THE ELLIPSE

OBJECTIVES

1. *Define the ellipse and derive its equation.*
2. *Given an equation of the ellipse, find the lengths of the major and minor axes, the foci, and the vertices.*
3. *Introduce translation of axes.*

The conic section that we shall discuss in this section is called the ellipse. We define the ellipse as follows:

DEFINITION 7.3 An *ellipse* is the set of all points in a plane the sum of whose distances from two fixed points is constant. The two fixed points are called the *foci* and the line passing through them is the *principal axis* of the ellipse.

We shall consider three special cases:

1. Foci on the x-axis with center at the origin.
2. Foci on the y-axis with center at the origin.
3. Center at $(h,k) \neq (0,0)$ with the principal axis parallel to one of the coordinate axes.

For case 1, we have

THEOREM 7.1   The ellipse with foci at $F_1(c,0)$ and $F_2(-c,0)$ is defined by

$$(x^2/a^2)+(y^2/b^2)=1 \qquad (1)$$

where $2a$ $(a>0)$ is the sum of the distances from any point on the ellipse to the two foci and $b=\sqrt{a^2-c^2}$.

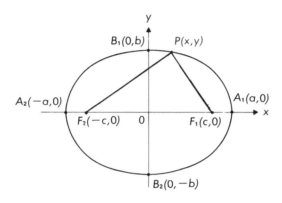

Figure 1

Proof. (See Figure 1). The point $P(x,y)$ belongs to the ellipse if and only if

$$d(P,F_1)+d(P,F_2)=2a \qquad (2)$$

Using the distance formula, we have

$$d(P,F_1)=\sqrt{(x-c)^2+(y-0)^2}$$

and

$$d(P,F_2)=\sqrt{(x+c)^2+(y-0)^2}$$

Thus we have the following relation describing the ellipse:

$$2a=\sqrt{(x-c)^2+y^2}+\sqrt{(x+c)^2+y^2} \qquad (3)$$

Performing the following algebraic manipulations, Equation 3 can be written as in Equation 1:

$$\sqrt{(x-c)^2+y^2}=2a-\sqrt{(x+c)^2+y^2}$$
$$((x-c)^2+y^2=4a^2-4a\sqrt{(x+c)^2+y^2}+(x+c)^2+y^2$$
$$a\sqrt{(x+c)^2+y^2}=a^2+cx$$
$$a^2[(x+c)^2+y^2]=a^4+2a^2cx+c^2x^2$$
$$a^2x^2+2a^2cx+a^2c^2+a^2y^2=a^4+2a^2cx+c^2x^2$$
$$(a^2-c^2)x^2+a^2y^2=a^2(a^2-c^2)$$
$$(x^2/a^2)+(y^2/b^2)=1$$

where $b^2=a^2-c^2$. This proves Theorem 7.1.

The points A_1 and A_2 are called the **vertices** of the ellipse; B_1 and B_2 are the y-intercepts. The line segment A_1A_2 is called the **major axis**; the line segment B_1B_2 is called the **minor axis**. In this case, the center of the ellipse is at the origin. Note that the length of the major axis is $2a$ and the length of the minor axis is $2b$.

EXAMPLE 1. Find the lengths of the major and minor axes and give the coordinates of the foci, vertices, and y-intercepts of the ellipse defined by

$$9x^2 + 16y^2 = 144 \tag{4}$$

Solution. Dividing Equation 4 by 144 and simplifying, we get

$$(x^2/16) + (y^2/9) = 1 \tag{5}$$

or

$$(x^2/4^2) + (y^2/3^2) = 1 \tag{6}$$

Comparing this with Equation 1, we find that $a = 4$ and $b = 3$. Hence,

$$\text{length of the major axis} = 2a = 8$$
$$\text{length of the minor axis} = 2b = 6$$

Since $b^2 = a^2 - c^2$, we find that

$$c = \sqrt{a^2 - b^2} = \sqrt{16 - 9} = \sqrt{7}$$

Therefore, the foci are $F_1(\sqrt{7}, 0)$ and $F_2(-\sqrt{7}, 0)$, and the vertices and y-intercepts are

$$A_1 = (a, 0) = (4, 0) \qquad\qquad B_1 = (0, b) = (0, 3)$$
$$A_2 = (-a, 0) = (-4, 0) \qquad\qquad B_2 = (0, -b) = (0, -3)$$

The graph is shown in Figure 2.

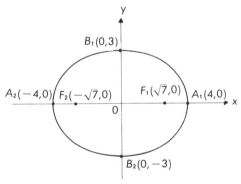

Figure 2

For case 2, we have the following result.

THEOREM 7.2   The ellipse with foci at $F_1(0, c)$ and $F_2(0, -c)$ is defined by

$$(x^2/b^2) + (y^2/a^2) = 1 \tag{7}$$

where $2a(a > 0)$ is the sum of the distances from any point on the ellipse to the two foci and

$$b = \sqrt{a^2 - c^2}$$

The proof will be left as an exercise.

In case 2, the major axis is along the y-axis and the minor axis is along the x-axis. The vertices are $A_1(0,a)$, $A_2(0,-a)$; the x-intercepts are $B_1(b,0)$ and $B_2(-b,0)$. This is illustrated in Figure 3.

EXAMPLE 2. Find an equation of the ellipse with foci at $F_1(0,4)$ and $F_2(0,-4)$ and the two vertices at $A_1(0,5)$ and $A_2(0,-5)$.

Solution. We note that $a=5$ and $c=4$. Hence $b=\sqrt{a^2-c^2}=\sqrt{25-16}=3$.

Since the foci are on the y-axis, we substitue these values into Equation 7 to obtain

$$(x^2/3^2)+(y^2/5^2)=1$$

or

$$(x^2/9)+(y^2/25)=1$$

The graph is shown in Figure 4.

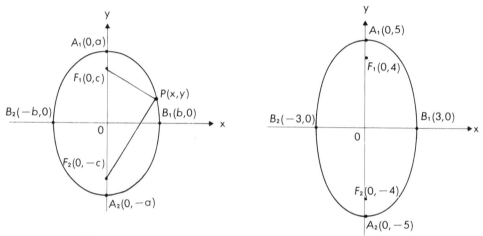

Figure 3

Figure 4

> ***Self-test.*** Consider the equation of an ellipse $(x^2/16)+(y^2/25)=1$.
>
> (a) The principal axis is along the ______-axis.
>
> (b) The vertices are at $A_1=$______ and $A_2=$______.
>
> (c) The x-intercepts are at $B_1=$______ and $B_2=$______.
>
> (d) The foci are at $F_1=$______ and $F_2=$______.

Before we consider case 3, it would be helpful to examine a concept known as the translation of axes. Consider two rectangular coordinate systems as shown in

ANSWERS: (a) y-axis (b) $(0,5)$, $(0,-5)$ (c) $(4,0)$, $(-4,0)$ (d) $(0,3)$, $(0,-3)$

Figure 5: the xy-system and the uv-system, where the u-axis is parallel to the x-axis. Let O' be the origin for the uv-system and suppose (h,k) are the coordinates of O' in the xy-system. Then, every point P in the plane has two sets of coordinates: P(x,y) in the xy-system, and P(u,v) in the uv-system. Clearly,

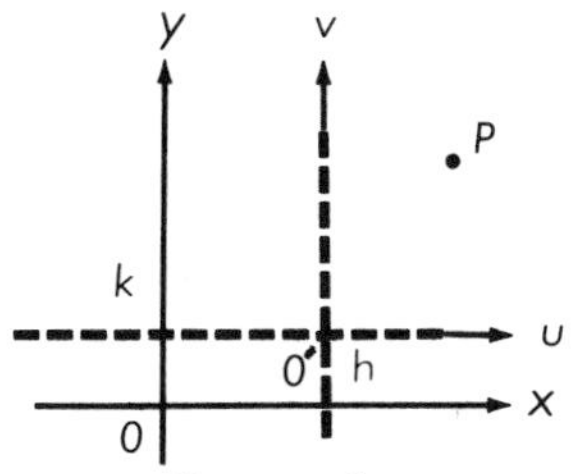

$$x = u + h \tag{8}$$
$$y = v + k$$

or

$$u = x - h$$
$$v = y - k \tag{9}$$

Figure 5

Equations 8 and 9 are called the equations for **translation of axes**. They relate the two sets of coordinate systems.

Now we can consider case 3. Here, the center of the ellipse is at the point (h,k) with the principal axis parallel to one of the coordinate axes. Let the principal axis be the line $y = k$ and the foci be at the points $F_1(h+c,k)$ and $F_2(h-c,k)$. Setting up the rectangular coordinate system, the uv-system with origin O' at (h,k) in the xy-system we find by Theorem 7.2 that the equation for the ellipse in the uv-system is

$$(u^2/a^2) + (v^2/b^2) = 1 \tag{10}$$

(See Figure 6.) Using Equations 9 for the translation of axes: $u = x - h$ and $v = y - k$. Equation 10 becomes:

$$[(x-h)^2/a^2] + [(y-k)^2/b^2] = 1 \tag{11}$$

which is the equation of an ellipse with center at (h,k), principal axis along the line $y = k$, major axis of length 2a, and minor axis of length 2b.

Similarly if the center of the ellipse is at (h,k), with principal axis along $x = h$ and major axis of length 2a (see Figure 7), then the equation of the ellipse is

$$[(x-h)^2/b^2] + [(y-k)^2/a^2] = 1 \tag{12}$$

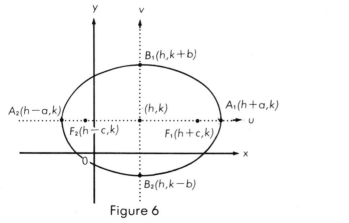

Figure 6

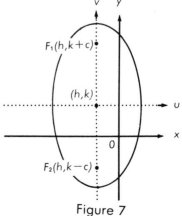

Figure 7

EXAMPLE 3. Show that the equation

$$9x^2 + 4y^2 - 36x + 8y + 4 = 0 \qquad (13)$$

defines an ellipse. Find the center, foci, and vertices.

Solution. First complete the square in x and y:

$$9x^2 - 36x + 4y^2 + 8y = -4$$
$$9(x^2 - 4x + 4) + 4(y^2 + 2y + 1) = -4 + 36 + 4$$

or

$$9(x - 2)^2 + 4(y + 1)^2 = 36$$

or

$$[(x-2)^2/4] + [(y+1)^2/9] = 1 \qquad (14)$$

Equation 14 is of the form given by Equation 12. Thus, Equation 13 defines an ellipse with its center at $(2, -1)$ and its principal axis as the line $x = 2$. Using the relation

$$b^2 = a^2 - c^2$$

we find

$$c = \sqrt{a^2 - b^2}$$

Here, $a^2 = 9$ and $b^2 = 4$. Therefore,

$$c = \sqrt{5}$$

and the foci are at the points $F_1(0, \sqrt{5})$ and $F_2(0, -\sqrt{5})$ on the u-axis or at $F_1(2, -1 + \sqrt{5})$ and $F_2(2, -1 - \sqrt{5})$ in the xy-system. The vertices are $A_1(2,2)$ and $A_2(2, -4)$. We also find $B_1(4, -1)$ and $B_2(0, -1)$. The ellipse is sketched in Figure 8.

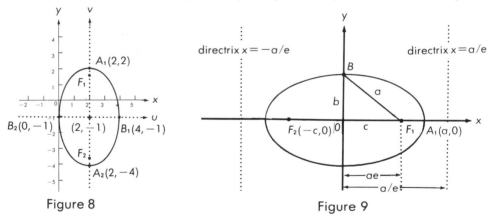

Figure 8   Figure 9

Geometrically, the essential feature of an ellipse is given by the equation

$$d(P,F_1) + d(P,F_2) = 2a \qquad (15)$$

where P is any point on the ellipse, F_1 and F_2 are the foci, and $a > 0$ is a constant.

Algebraically, the essential property of an ellipse is that its equation in quadratic form without a cross-product term in x and y, involves x^2 and y^2 terms having the same sign. When looking at the properties of the ellipse, it is helpful to look at the equation in the reduced form

$$(x^2/a^2)+(y^2/b^2)=1, \quad a>b>0 \tag{16}$$

where $b^2=a^2-c^2$ and $F_1(c,0)$ and $F_2(-c,0)$ are the two foci of the ellipse and $2a$ is the length of the major axis (see Figure 9).

We note that if B is a y-intercept, then

$$d(B,F_1)=d(B,F_2)$$

and since (from Equation 15)

$$d(B,F_1)+d(B,F_2)=2a$$

we have

$$d(B,F_1)=d(B,F_2)=a$$

The eccentricity e of an ellipse is defined by

$$e=c/a \tag{17}$$

Since $c<a$, we have $0 \le e<1$. The eccentricity indicates the degree of flatness of the ellipse. If a is kept fixed and $c=0$, then $e=0$ and we have $F_1=F_2$ and the resulting conic is a circle. As c increases, e increases and the ellipse becomes flatter until in the extreme case, $c=a$ and $e=1$, the "ellipse" reduces to the line segment F_1F_2.

From Definition 7.1, we have

$$d(B,F_1)=ed(B,D_1)$$

where $d(B,D_1)$ is the distance from B to the directrix associated with $F_1(c,0)$. Since

$$d(B,F_1)=a \text{ then } d(B,D_1)=a/e$$

Thus, the directrix associated with F_1 is the line $x=a/e$. Similarly, the directrix associated with $F_2(-c,0)$ is the line $x=-a/e$.

In general, the equations of an ellipse are

$$d(P,F_1)=ed(P,D_1)$$
$$d(P,F_2)=ed(P,D_2) \tag{18}$$

where e is the eccentricity, P is any point on the ellipse, and F_1 and F_2 are the foci. The distance $d(P,D_1)$ is measured from P to the directrix at the same end of the ellipse as F_1 and $d(P,D_2)$ is the distance measured from P to the directrix at the same end of the ellipse as F_2.

EXERCISES 7.3

In Exercises 1-12, find the length of the major and minor axes, the foci, vertices, and eccentricity. Sketch the graph.

1. $16x^2 + 9y^2 = 144$ 2. $25x^2 + 16y^2 = 400$ 3. $16x^2 + 25y^2 = 400$

4. $4x^2 + 3y^2 = 12$ 5. $5x^2 + 2y^2 = 20$ 6. $3x^2 + 2y^2 = 10$

7. $3x^2 + y^2 = 9$ 8. $9x^2 + 4y^2 = 1$ 9. $(x-1)^2 + 4(y-3)^2 = 16$

10. $16(x+2)^2 + 9(y-3)^2 = 144$ 11. $3x^2 + 2y^2 - 6x + 4y + 5 = 0$

12. $4x^2 + 16y^2 + 24x + 20 = 0$

In Exercises 13-25, find an equation of an ellipse which satisfies the given conditions and sketch the graph.

13. Center at the origin, the length of the major axis is 8, the length of the minor axis is 6, and the foci are on the y-axis.

14. Foci at (3,0) and (−3,0), vertices at (5,0) and (−5,0).

15. Foci at (0,4) and (0, −4), vertices at (0,5) and (0, −5).

16. Center at the origin, one focus at the point $(0, \sqrt{5})$, and the length of the major axis is 8.

17. Vertices at (5,0) and (−5,0) and contains the point (3,3).

18. Foci at (3,0) and (−3,0) and contains the point (4, −1).

19. Foci at (−1,1) and (1,1) and passing through the origin.

20. Vertices at (4,0) and (−4,0), eccentricity $e = \frac{4}{5}$.

21. Foci at (4,0) and (−4,0), eccentricity $e = \frac{4}{5}$.

22. Foci along the x-axis, center at the origin, eccentricity $e = \frac{3}{4}$, and passing through the point (6,4).

23. Foci at (4,0) and (−4,0), directrices the lines $x = 6$ and $x = −6$.

24. Eccentricity $e = \frac{2}{3}$, directrices the lines $y = 6$ and $y = −6$.

25. Eccentricity $e = \frac{1}{2}$, directrices the lines $y = 7$ and $y = −7$.

26. Show that the length of the latus rectum of the ellipse $b^2x^2+a^2y^2=a^2b^2$ $(a>b>0)$ is $2b^2/a$.

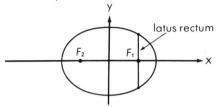

27. The earth travels around the sun in an elliptical orbit, with the sun at one focus. If the major axis is 185.8 million miles long and the eccentricity of the orbit is 0.017, find:
    (a) The shortest distance between the earth and the sun, which occurs when the earth is at the perigee.
    (b) The greatest distance between the earth and the sun, which occurs when the earth is at the apogee.

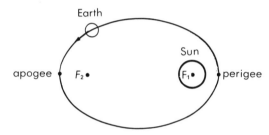

28. An earth satellite is in an elliptic orbit. If the length of the major axis is 10^4 miles and the length of the minor axis is 9,500 miles, and the radius of the earth is assumed to be 4,000 miles, find:
    (a) The distance of the perigee above the earth's surface.
    (b) The distance of the apogee above the earth's surface.

29. An earth satellite has an apogee of 250,000 miles and a perigee of 25,000 miles measured from the surface of the earth. What is the eccentricity of the elliptical orbit of the satellite?

30. The orbit of a satellite has eccentricity of 0.75. If the apogee is 26,500 miles, find its perigee.

7.4 THE PARABOLA

In this section, we shall examine another of the conic sections, the parabola.

> **DEFINITION 7.4** A *parabola* is the set of all points in a plane equidistant from a fixed line and a fixed point not on the line. The fixed line is called the *directrix* and the fixed point is called the *focus*.

We shall consider two special cases concerning parabolas.

1. We shall assume that the focus is on the x-axis at $(p,0)$ and the directrix is defined by the line $x = -p$.
2. We shall assume that the focus is on the y-axis at $(0,p)$ and the directrix is the line defined by $y = -p$.

THEOREM 7.3 The parabola with focus at the point $(p,0)$ and the directrix the line $x = -p$ is defined by

$$y^2 = 4px \qquad (1)$$

Proof. The point $P(x,y)$ belongs to the parabola if and only if the distance from P to the focus $F(p,0)$ is equal to the distance from P to the directrix. From P draw a line perpendicular to the directrix (see Figure 1) intersecting it at the point Q. Then, the coordinates of Q are $(-p,y)$. Thus we have

$$d(P,Q) = d(P,F)$$

and using the distance formula, we obtain

$$\sqrt{(x+p)^2+(y-y)^2}=\sqrt{(x-p)^2+(y-0)^2}$$
$$x+p=\sqrt{(x-p)^2+y^2}$$

Therefore, by squaring both sides of the equation, we get

$$(x+p)^2=(x-p)^2+y^2$$

or

$$x^2+2px+p^2=x^2-2px+p^2+y^2$$

Simplifying we obtain

$$y^2=4px$$

This proves Theorem 7.3.

   In Figure 1, $p>0$. If $p<0$, we obtain the graph shown in Figure 2. In each case, the x-axis is the axis of the parabola (the line of symmetry) and the origin, which is halfway between the focus and the directrix, is the vertex of the parabola.

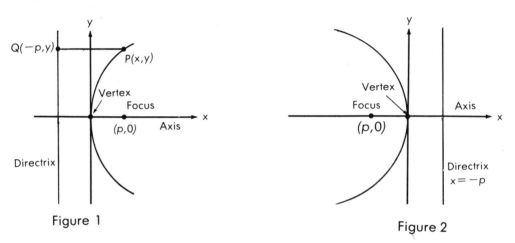

Figure 1   Figure 2

EXAMPLE 1. Find the equation of the parabola with focus at the point (3,0) and directrix $x=-3$.

Solution. Here $p=3$. Applying Theorem 7.3, we have

$$y^2=9x$$

The graph is shown in Figure 3.

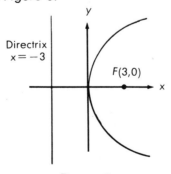

Figure 3

EXAMPLE 2. Find the directrix and the coordinates of the focus of the parabola whose equation is

$$y^2 = -7x \tag{2}$$

Solution. Equation 2 may be written as

$$y^2 = 4(-7/4)x \tag{3}$$

Comparing this with Equation 1, we find that $p = -\frac{7}{4}$. Therefore, the focus is the point $(-\frac{7}{4}, 0)$ and the directrix is $x = -(-\frac{7}{4}) = \frac{7}{4}$. The graph is shown in Figure 4.

We now return to special case 2: The focus is on the y-axis at $(0,p)$ and the directrix is the line $y = -p$. The following theorem can be easily proved.

THEOREM 7.4 The parabola with focus at the point $(0,p)$ and directrix the line $y = -p$ is defined by

$$x^2 = 4py \tag{4}$$

See Figure 5. The reader will notice that Equation 4 is the equation of a quadratic function.

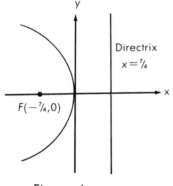

Figure 4

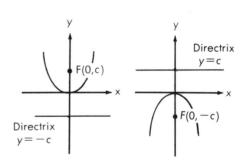

Figure 5

EXAMPLE 3. Find the equation of a parabola containing the point $(2, -3)$ with its vertex at the origin and whose axis is the y-axis.

Solution. Since the vertex is at the origin and the axis of the parabola is the y-axis, the parabola is the graph of a quadratic function defined by

$$x^2 = 4py \tag{5}$$

The point $(2, -3)$ belongs to this function. Hence, substitution into Equation 5 yields

$$y = (1/4p)x^2$$
$$-3 = (1/4p)(2^2)$$

or

$$-3 = 1/p$$

Therefore, the required equation is

$$y = \tfrac{1}{4}(-3)x^2$$

or

$$y = -(\tfrac{3}{4})x^2$$

Since $p = -\tfrac{1}{3}$, the focus is at $(-\tfrac{1}{3},0)$ and the directrix is the line $y = \tfrac{1}{3}$ (see Figure 6).

Self-test. Consider the equation of the parabola $x = -y^2$.

    (a) The focus of the parabola is at _____.

    (b) The directrix is the line $x =$ _____.

If the vertex of a parabola is at a point other than the origin, Equations 1 and 4 no longer apply. However, it is a simple matter to use Definition 7.4 to determine the appropriate equation.

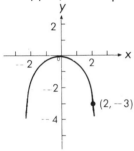

Figure 6

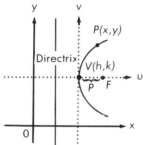

Figure 7

Suppose we consider a parabola with its vertex at $V(h,k)$ and the principal axis the line $y = k$. In this case (see Figure 7) the focus is at the point $F(h+p,k)$ and the directrix is the line $x = h-p$ where p is positive. Setting up a rectangular coordinate system, the uv-system, with its origin at (h,k) we find that

$$v^2 = 4pu \qquad (6)$$

Using the equations for translation of axes, we obtain

$$(y-k)^2 = 4p(x-h) \qquad (7)$$

For the case where the parabola has its vertex at $V(h,k)$ and focus at $F(h-p,k)$ we obtain

$$(y-k)^2 = -4p(x-h) \qquad (8)$$

Similarly, if the vertex of the parabola is at $V(h,k)$ and the focus is at $F(h,k+p)$ then the equation of the parabola is given by

$$(x-h)^2 = 4p(y-k) \qquad (9)$$

while if the vertex is at $V(h,k)$ and the focus is at $F(h,k-p)$ we obtain

$$(x-h)^2 = -4p(y-k) \qquad (10)$$

ANSWERS:

EXAMPLE 4. Find an equation of the parabola with focus at $F(2,3)$ and its directrix the line $x=-3$.

Solution. (See Figure 8). Since the vertex V must be equidistant from the directrix and the focus, we have

$$p=(2+3)/2=\tfrac{5}{2}$$

Thus, the vertex is at $V(-\tfrac{1}{2},3)$. From Equation 7 we obtain

$$(y-3)^2=4(\tfrac{5}{2})[x-(-\tfrac{1}{2})]$$

or

$$(y-3)^2=10(x+\tfrac{1}{2})$$

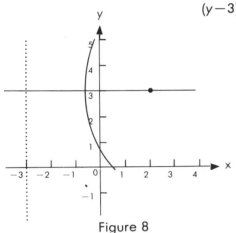

Figure 8

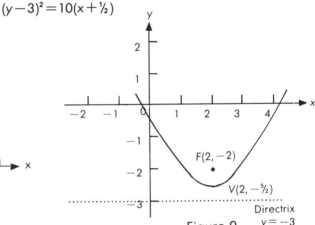

Figure 9

EXAMPLE 5. Given the equation of a parabola

$$x^2-4x-2y=1$$

find the vertex, focus, and directrix. Sketch the graph.

Solution. Completing the square in x we have

$$x^2-4x+4=2y+1+4$$
$$(x-2)^2=2y+5$$
$$=2(y+\tfrac{5}{2})$$

Thus, $p=\tfrac{1}{2}$, $h=2$, and $k=-\tfrac{5}{2}$. The vertex is at $V(2,-\tfrac{5}{2})$. The focus is at $F(2,-2)$ and the directrix is the line $y=-3$. A sketch of the curve is shown in Figure 9.

Self-test. Consider the equation of a parabola $x^2+6x-8y+17=0$.
  (a)   The vertex is at the point $V=$ _____.
  (b)   The focus is at the point $F=$ _____
  (c)   The directrix is the line $y=$ _____

EXERCISES 7.4

In Exercises 1-10, find an equation of a parabola with its vertex at the origin which satisfies the given conditions. Sketch the graph, showing the vertex, focus, and directrix.

1. Focus at $(-2,0)$ 2. Focus at $(0,3)$ 3. Directrix: $y=4$

4. Directrix: $x=-\frac{2}{3}$ 5. Directrix: $x=\frac{5}{2}$ 6. Directrix: $y=-3$

7. Contains the point $(-3,6)$ and its principal axis is the x-axis.

8. Contains the point $(-3,6)$ and its principal axis is the y-axis.

9. Contains the point $(2,4)$ and its principal axis is the x-axis.

10. Opens upward and passes through the point $(-3,7)$.

In Exercises 11-18, use Definition 7.4 to find an equation of the parabola, given the focus F and the directrix. In each case, locate the vertex and the principal axis of the parabola.

11. $F(0,1)$, directrix: $y=-5$ 12. $F(-1,0)$, directrix: $y=2$

13. $F(5,2)$, directrix: $x=-1$ 14. $F(2,3)$, directrix: $y=-1$

15. $F(2,3)$, directrix: $y=5$ 16. $F(-1,2)$, directrix: $x=2$

17. $F(1,1)$, directrix: $x=-4$ 18. $F(1,-4)$, directrix: $x=5$

In Exercises 19-26, find the vertex, focus, and directrix of the given parabolas. Sketch the graph.

19. $x^2-2x+8y=6$ 20. $x^2-2y+8x=-10$

21. $y^2+4x-8=0$ 22. $x^2+4y+3x=2$

23. $4x^2+40x+y+106=0$ 24. $x^2+20y=10$

25. $y^2-4y+2x+7=0$ 26. $3y^2-8x-12y=4$

In Exercises 27-30, find an equation of the parabola after a translation of axes to the new origin indicated.

27. $x^2-6y+9=0$, $(0,\frac{3}{2})$ 28. $y^2+10y+6x+19=0$, $(1,-5)$

29. $x^2+6x-12y+57=0$, $(-3,4)$ 30. $y^2+8y+6x+1=0$, $(-3,4)$

7.5 THE HYPERBOLA

OBJECTIVES

1. Define a hyperbola and derive its equation.
2. Given an equation of a hyperbola find the vertices, foci, eccentricity, and asymptotes.

The hyperbola is another conic section with two foci and a center. We define the hyperbola as follows:

DEFINITION 7.5 A *hyperbola* is the set of all points in a plane the difference of whose distances from two fixed points is constant. The two fixed points are called the *foci* of the hyperbola.

Suppose the distance between the two foci is $2c$, $c > 0$. To simplify our discussion, we place the foci at the points $F_1(c,0)$ and $F_2(-c,0)$. Thus the center of the hyperbola is at the origin and the principal axis is the x-axis. Now, let $P(x,y)$ be any point on the hyperbola (see Figure 1), then by Definition 7.5, we have

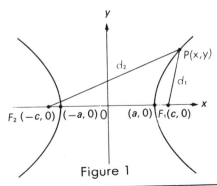

Figure 1

NOTE: In triangle PF_2F_3, (see Figure 1 above) $2c + d_1 > d_2$ and $2c + d_2 > d_1$. Therefore, $2c > |d_1 - d_2|$. But for the hyperbola $|d_1 - d_2| = 2a$, and so, $2c > 2a$.

$$d(P,F_1) - d(P,F_2) = 2a \tag{1}$$

or

$$d(P,F_2) - d(P,F_1) = 2a \tag{2}$$

Combining Equations 1 and 2, we may write

$$d(P,F_1) - d(P,F_2) = \pm 2a \tag{3}$$

Using the distance formula, then squaring and simplifying we obtain

$$(x^2/a^2) - [y^2/(c^2 - a^2)] = 1 \tag{4}$$

Since $2c > 2a$ (see note at bottom of page), then $c > a$ and $c^2 - a^2 > 0$. We define

$$b^2 = c^2 - a^2 \tag{5}$$

Then Equation 4 becomes

$$(x^2/a^2) - (y^2/b^2) = 1 \tag{6}$$

Equation 6 is an equation of a hyperbola. It is similar to the equation of an ellipse with the exception of the minus sign and the relationship between a, b, and c given in Equation 5.

It can be shown that the steps involved in going from Equation 3 to Equation 6, are reversible, which means that every point which satisfies Equation 6, is a point on the hyperbola.

The following observations can be made regarding hyperbolas:

1. From Equation 6, it is clear that the hyperbola is symmetric with respect to both the x-axis and the y-axis.

2. If P is a point on the hyperbola to the right of the line $x = a$, then the condition $d(P,F_1) - d(P,F_2) = 2a$ holds. If P is to the left of the line $x = -a$, then the condition $d(P,F_2) - d(P,F_1) = 2a$ holds. If P is between the lines $x = a$ and $x = -a$, then P is not on the hyperbola.

3. The hyperbola defined by Equation 6 has, as its asymptotes the lines
    $$y = (b/a)x \quad \text{and} \quad y = -(b/a)x$$

4. The length of $2b$ is a useful quantity. A rectangle constructed from the points $(a,0)$, $(-a,0)$, $(0,b)$, and $(0,-b)$ is called the **central rectangle**. The asymptotes of the hyperbola contain the diagonals of the central rectangle. The length of each diagonal is $2c$. This information is helpful when it comes to making a sketch of the graph of a hyperbola.

The line passing through the vertices $V_1(a,0)$ and $V_2(-a,0)$ is called the **transverse axis**; the line passing through the center perpendicular to the transverse axis is called the **conjugate axis**. The length of the transverse axis is $2a$, while the length of the conjugate axis is $2b$ (see Figure 2).

The **eccentricity** e of a hyperbola is defined by

$$e = c/a \tag{7}$$

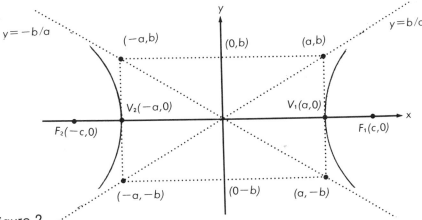

Figure 2

Let $P(x,y)$ be any point on the right branch of the hyperbola. We find for $D_1(a/e,y)$ and $F_1(c,0)$ that (see Figure 3)

$$d(P,F_1)=\sqrt{(x-c)^2+y^2}$$

$$=\sqrt{(x-c)^2+(b^2/a^2)x-b^2}$$

$$=\sqrt{x^2[(b^2+a^2)/a^2]-2cx+c^2-b^2}$$

$$=\sqrt{(c^2/a^2)x^2-2cx+a^2}$$

$$=\sqrt{[(c/a)x-a]^2}$$

$$=|(c/a)x-a|$$

$$=c/a|x-(a^2/c)|=e|x-(a/e)|$$

But

$$d(P,D_1)=\sqrt{[x-(a/e)]^2+(y-y)^2}$$
$$=x-(a/e)$$

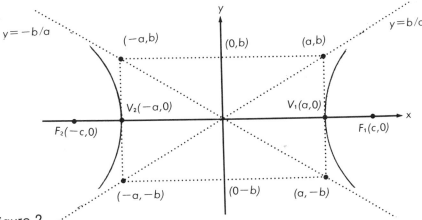

Figure 3

Therefore,

$$d(P,F_1)=ed(P,D_1)$$ (8)

Similarly, it can be shown that

$$d(P,F_2)=ed(P,D_2)$$ (9)

From Equations 8 and 9, we obtain

$$d(P,F_2)-d(P,F_1)=e[d(P,D_2)-d(P,D_1)]$$
$$=e[(x+a/e)-(x-a/e)]=2a$$

as stated in Definition 7.5.

If the foci of the hyperbola are $F_1(0,c)$ and $F_2(0,-c)$ the equation is of the form

$$(y^2/a^2)-(x^2/b^2)=1$$ (10)

where

$$b^2=c^2-a^2$$ (11)

See Figure 4.

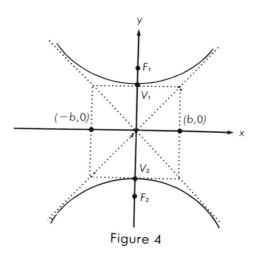

Figure 4

From Equations 6 and 10, we see that the relative size of a and b plays no part in determining where the foci and axes are. This is in contrast with the equation of the ellipse. Thus, for Equation 6, the transverse axis is always on the x-axis and for Equation 10, the transverse axis is always on the y-axis, which is the case whether $a>b>0$ or $b>a>0$.

EXAMPLE 1. Given the hyperbola with the equation

$$16x^2-25y^2=400$$ (12)

find the vertices, foci, and eccentricity. Sketch the graph.

Solution. Dividing Equation 12 by 400, we obtain

$$(x^2/25)-(y^2/16)=1$$ (13)

which is of the form given by Equation 6. Thus, the transverse axis is along the x-axis with the center of the hyperbola at the origin. Here,

$$a=5, \text{ and } b=4$$

The vertices are at $V_1(5,0)$ and $V_2(-5,0)$. From
$$c^2 = a^2 + b^2$$
we have
$$c = \pm\sqrt{5^2 + 4^2} = \pm\sqrt{41}$$
Therefore, the foci are at $F_1(\sqrt{41},0)$ and $F_2(-\sqrt{41},0)$. The eccentricity is
$$e = c/a = \sqrt{41}/5$$
We construct the central rectangle and draw the asymptotes:
$$y = \tfrac{4}{5}x \text{ and } y = -\tfrac{4}{5}x$$
A sketch of the graph is shown in Figure 5.

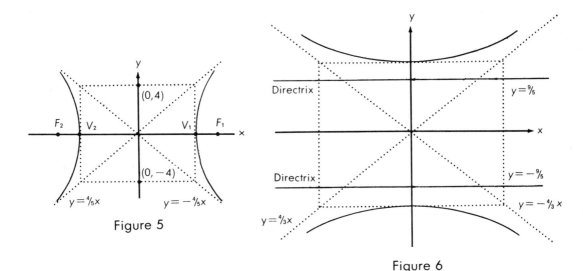

Figure 5

Figure 6

EXAMPLE 2. Given the hyperbola with equation
$$16y^2 - 9x^2 = 144 \tag{14}$$
find the foci, vertices, eccentricity, directrices, and asymptotes. Draw a sketch of the graph.

Solution. Dividing Equation 14 by 144 we get
$$(y^2/9) - (x^2/16) = 1 \tag{15}$$
which is of the form given by Equation 10. Here, $a = 3$ and $b = 4$. The vertices are at the points $V_1(0,3)$ and $V_2(0,-3)$. Since, $c = \pm\sqrt{9+16} = \pm 5$, the foci are at the points $F_1(0,5)$ and $F_2 = (0,-5)$. The eccentricity $e = \tfrac{5}{3}$ and so the directrices are the lines $y = \pm a/e$ or
$$y = \tfrac{9}{5} \text{ and } y = -\tfrac{9}{5}$$
The asymptotes are the lines $y = \pm(b/a)x$
or
$$y = \tfrac{4}{3}x \text{ and } y = -\tfrac{4}{3}x$$
A sketch of the graph is shown in Figure 6.

If the center of the hyperbola is at the point $(h,k) \neq (0,0)$ and the transverse axis is parallel to the x-axis, then a translation of axes yields the equation

$$[(x-h)^2/a^2] - [(y-k)^2/b^2] = 1 \qquad (16)$$

If the center of the hyperbola is at the point $(h,k) \neq (0,0)$ and the transverse axis is parallel to the y-axis, then a translation of axes yields the equation

$$[(y-k)^2/a^2] - [(x-h)^2/b^2] = 1 \qquad (17)$$

EXAMPLE 3. Show that the equation

$$4x^2 - 9y^2 - 16x - 54y = 101 \qquad (18)$$

defines a hyperbola. Find the center, foci, vertices, eccentricity, and asymptotes.

Solution. We complete the squares in x and y:

$$4x^2 - 9y^2 - 16x - 54y = 101$$
$$4(x^2 - 4x + 4) - 9(y^2 + 6y + 9) = 101 + 16 - 81$$

or

$$4(x-2)^2 - 9(y+3)^2 = 36$$

or

$$[(x-2)^2/9] - [(y+3)^2/4] = 1 \qquad (19)$$

Thus, Equation 18 is in the form of Equation 16. The center of the hyperbola is the point $(2, -3)$ and the transverse axis is along the line $y = -3$. Here $a = 3$ and $b = 2$. Therefore, the vertices are at the points $V_1(5, -3)$ and $V_2(-1, -3)$. Since $c = \sqrt{9+4} = \sqrt{13}$, then the foci are at the points $F_1(2+\sqrt{13}, -3)$ and $F_2(2-\sqrt{13}, -3)$. The eccentricity $e = c/a = \sqrt{13}/3$ and the asymptotes have slopes $m = \pm b/a = \pm\frac{2}{3}$ and pass through the center $(2, -3)$. Therefore, the equations for the asymptotes are

$$y + 3 = \frac{2}{3}(x-2) \qquad (20)$$

and

$$y + 3 = -\frac{2}{3}(x-2) \qquad (21)$$

A sketch of the graph is shown in Figure 7.

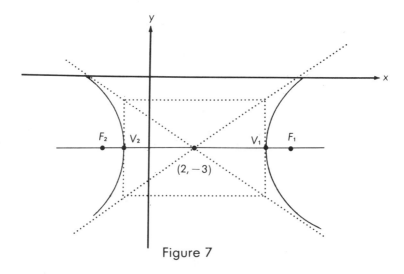

Figure 7

EXERCISES 7.5

In Exercises 1-12, find the center, foci, eccentricity, vertices, and lengths of the transverse and conjugate axes of the given equation. Find the equation of the asymptotes and sketch the graph.

1. $9x^2 - 16y^2 = 144$ 2. $(x^2/9) - (y^2/16) = 1$ 3. $9y^2 - 16x^2 = 144$

4. $(y^2/4) - (x^2/9) = 1$ 5. $3x^2 - 2y^2 = 10$ 6. $5x^2 - 3y^2 = 1$

7. $9(x+1)^2 - 4(y-1)^2 = 36$ 8. $4(y-1)^2 - 9(x+1)^2 = 36$

9. $x^2 - 4y^2 - 2x - 16y = 20$ 10. $y^2 - 4x^2 - 8x - 2y = 7$

11. $9x^2 - 4y^2 - 18x - 16y + 29 = 0$ 12. $y^2 - x^2 + 2x - 2y = 1$

In Exercises 13-22, find an equation of a hyperbola which satisfies the given conditions and sketch the graph.

13. Foci at $(4,0)$ and $(-4,0)$, vertices at $(3,0)$ and $(-3,0)$.

14. Foci at $(0,4)$ and $(0,-4)$, eccentricity $e = \frac{3}{2}$.

15. Center at $(-1,1)$, a focus at $(2,1)$, and a vertex at $(-1+\sqrt{5},1)$.

16. Foci at $(6,0)$ and $(-6,0)$ and asymptotes the lines $y = \pm\frac{4}{3}x$.

17. Center at $(-5,-2)$, vertex at $(-5,0)$ and asymptotes the lines $y + 2 = \pm\frac{1}{2}(x+5)$.

18. Vertices at $(0,4)$ and $(0,-4)$ and asymptotes the lines $y = \pm 2x$.

19. Center at the origin, the transverse axis along the y-axis, and passing through the points $(2,4)$ and $(7,-6)$.

20. Vertices at $(-5,-3)$ and $(-5,-1)$, eccentricity $e = \sqrt{5}$.

21. Center at $(1,-2)$, one directrix the line $y = -2 + (9/\sqrt{13})$, and a focus at $(1,1)$.

22. Foci at $(8,0)$ and $(-8,0)$ and a conjugate axis with a length of 6.

7.6 POLAR EQUATIONS OF CONICS

OBJECTIVES

1. Derive polar equations of conics.
2. Identify and sketch the graph of conics described in polar form.

It is often convenient to describe conics in a polar coordinate system. We now derive a polar equation of a conic, defined in Section 7.1 as a set of points P such that the ratio of the distance between the focus F and P to the distance between the directrix and P is a positive constant.

Let a focus F of the conic coincide with the pole, and let the directrix be perpendicular to the polar axis. If $P(r,\theta)$ is a point on the conic, then by Definition 7.1 we have (see Figure 1):

$$d(P,F)/d(P,D)=e \tag{1}$$

or

$$\frac{r}{p+r\cos\theta}=e$$

Solving for r, we get

$$r=\frac{ep}{(1-e\cos\theta)} \tag{2}$$

Conversely, if r and θ satisfy Equation 2, the point (r,θ) is on a conic with eccentricity e and with directrix and focus as shown in Figure 1.

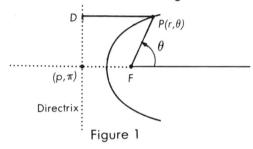

Figure 1

If the directrix is to the right of the pole a distance p, then the equation of the conic becomes

$$r=ep/(1+e\cos\theta) \tag{3}$$

If the directrix is parallel to the polar axis, the conic has equation

$$r=ep/(1\pm e\sin\theta) \tag{4}$$

depending on whether the directrix is above or below the polar axis.

EXAMPLE 1. Sketch the graph of $r=4/(2-\cos\theta)$.

Solution. We put this equation in the form of Equation 2 and get

$$r=\tfrac{1}{2}(4)/(1-\tfrac{1}{2}\cos\theta)$$

We see that the curve has eccentricity $e=\tfrac{1}{2}$ and $p=4$. Thus, the directrix is 4 units to the left of the pole and the conic is an ellipse as shown in Figure 2.

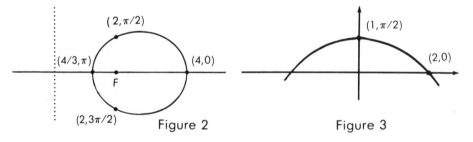

Figure 2   Figure 3

Self-test. Let $r=3/(1+2\cos\theta)$. Then

    (a) $e=$ _____ (b) $p=$ _____ (c) The conic is a _____

EXAMPLE 2. Sketch the graph of $r=2/(1+\sin\theta)$.

Solution. We see that the curve has eccentricity $e=1$ and so the conic is a parabola. Here, $p=2$. The graph is sketched in Figure 3.

EXERCISES 7.6

In Exercises 1-12, identify and sketch the graph of the given equation.

1. $r=6/(3+\sin\theta)$ 2. $r=6/(1+3\cos\theta)$ 3. $r=6/(2-6\cos\theta)$

4. $r=10/(3+2\cos\theta)$ 5. $r=10/(3-2\cos\theta)$ 6. $r=10/(3-2\sin\theta)$

7. $r=10/(2+3\sin\theta)$ 8. $r=3/(2-2\sin\theta)$ 9. $r=5/(2+3\cos\theta)$

10. $r=(3\csc\theta)/(2\csc\theta-2)$ 11. $r=\sec\theta(\sec\theta-\tan\theta)$

12. $2r\csc\theta+3r=6\csc\theta$

13. Find a polar equation of the parabola with focus at the pole and and vertex at $(2, \pi/2)$.

14. Find a polar equation of the parabola with focus at the pole and vertex at $(4, \pi)$.

15. Find a polar equation of the ellipse with eccentricity $\frac{1}{3}$, a vertex at $(1, \pi/2)$, and a focus at the pole.

16. Find a polar equation of the conic with eccentricity $\frac{2}{3}$, a vertex at $(1, \pi)$, and a focus at the pole.

17. Derive the formula $r=ep/(1\pm e\cos\theta)$.

18. Derive the formula $r=ep/(1\pm e\sin\theta)$.

REVIEW OF CHAPTER SEVEN

1. Define a conic section.

2. Define a directrix.

3. Define the eccentricity of a conic.

4. Define a circle.

5. The equation of a circle with its center at $(2, -3)$ and radius of 4 is
    _____.

6. Consider an equation of a circle which is $x^2 + y^2 + 4x - 6y - 12 = 0$. The
    center is at the point $C =$ _____ and the radius $r =$ _____.

7. Define an ellipse.

8. Consider an ellipse with the equation $9x^2 + 4y^2 = 36$. The vertices are
    $A_1 =$ _____ and $A_2 =$ _____.
    The x-intercepts are _____ and _____.

9. If the major axis of an ellipse has a length of 6 and the y-intercepts
    are 2 and -2, then the equation of the ellipse is _____.

10. If the minor axis of an ellipse has a length of 12 and the foci are at
    $(4,0)$ and $(-4,0)$, then the equation of the ellipse is _____.

11. Consider an ellipse with the equation $x^2 + 4y^2 - 4x + 8y + 4 = 0$. The
    center is at $C =$ _____, the foci are at _____,
    and _____, and the length of the major axis is _____.

12. Define a parabola.

13. How many foci does a parabola have?

14. Derive an equation of a parabola with focus $F(4,0)$ and directrix the
    line $x = -2$.

15. Given a parabola with the equation $x = \frac{1}{8}(y-2)^2 + 1$, the focus is at
    $F =$ _____, the vertex is the point $V =$ _____,
    and the directrix is the line $x =$ _____.

16. Define a hyperbola.

17. How many foci does a hyperbola have?

18. If a hyperbola has the equation $(x^2/a^2)-(y^2/b^2)=1$, then the foci are $F_1 = $ _____ and $F_2 = $ _____.
The equations for the asymptotes are $y = $ _____ and $y = $ _____.

19. Find an equation of the hyperbola with foci at $(5,0)$ and $(-5,0)$ and a conjugate axis with a length of 6.

20. Find an equation of a hyperbola with its center at the origin, length of the transverse axis 10, and length of the conjugate axis 4.

21. Show that a conic section in polar form is given by the equation $r = pe/(1 \pm e \sin \theta)$ if the directrix is parallel to the polar axis and a focus is at the pole.

22. In the equation above, if $e = 1$, the conic is a _____;
if $0 < e < 1$, the conic is a _____; if $e > 1$, the conic is a _____.

TEST SEVEN

1. Find the center and radius of the circle defined by
$$x^2+y^2-6x+4y-4=0$$

2. Find an equation of a circle with its center at (3,4) and tangent to the line $x=-1$.

3. Consider the equation of an ellipse
$$x^2+3y^2=9$$
   Find:
          (a)   The length of the major and minor axes.
          (b)   The foci.
          (c)   The vertices.
          (d)   The eccentricity.
   Sketch the graph.

4. Find an equation of an ellipse with foci at (0,3) and (0, −3) and containing the point (−1,4).

5. Find an equation of a parabola with focus at (2,5) and directrix $y=-1$.

6. Consider the equation of a parabola
$$4y^2+40y+x+106=0$$
   Find:
          (a)   The vertex.
          (b)   The focus.
          (c)   The directrix.
   Sketch the graph.

7. Find an equation of a hyperbola with its center at (1, −1) and a vertex at $(1,-1+\sqrt{5})$.

8. Given the following equation of a conic:
$$3x^2-16x-y^2+16=0$$
   Find:
          (a)   The polar equation of the conic.
          (b)   The eccentricity.
          (c)   The directrix.

9. Identify and sketch the graph of the equation:
$$r=12/(3-2\cos\theta).$$

APPENDIX A:
REVIEW OF SETS AND
LOGARITHMS

A.1 SETS

OBJECTIVES

1. *Introduce the concept of a set and discuss universal sets, subsets, and equality of sets.*
2. *Discuss the intersection, union, and complement of sets.*

The concept of a **set** or collection of objects is very natural to our thought processes. In everyday life we all encounter sets. For example:

1. The set of students at your college.
2. The set of continents on the earth.
3. The set of ballet dancers in a performance of *Sleeping Beauty*.
4. The set of all women presidents of the United States prior to 1980.
5. The set of all counting numbers.
6. The set of all known suns in our solar system.

The term set is undefined; the objects in a set are referred to as **elements** of the set. We shall assume that a collection of objects that is **well-defined** is a set. This means that if S is a set and x is an object, then it must be unambiguously clear whether or not x belongs to S. Thus, "the collection of all good teachers at your college" is ill-defined. The statement does not tell us what a good teacher is. On the other hand, "all mathematics instructors at your college" is a well-defined collection and therefore is a set.

A set may have one element, as in Example 6, or no element, as in Example 4. A set may have more than one, but a finite number of elements, such as Examples 1,2, and 3, or a set may have an infinite number of elements as in Example 5. The set described in 5 is called an **infinite** set, whereas the other sets are called **finite**. The set which has no element is called an **empty** set and is denoted by the symbol $\emptyset$.

Capital letters such as $A,B,C,X,Y...$ are used to denote sets. Lower case letters such as $a,b,x,y...$ are used to denote elements of a set. We use the notation

$$x \in S$$

read "x is an element of the set S". If y is not an element of the set S, we write

$$y \notin S$$

There are two general methods for describing sets. First, a set may be described by listing all of the elements within braces and then separating by commas. For example, if S is the set consisting of the first three counting numbers, we write

$$S = \{1,2,3\}$$

If there are a large number of elements in a set, it is cumbersome to list each element. Thus, we can write

$$A = \{2,4,6,8,...\}$$

to show the set of all even numbers equal to or greater than 2.

The second method of designating a set consists of describing an identifying property of a set. If B is a set consisting of elements x having property p, we write

$$B = \{x \mid x \text{ has property } p\}$$

which is read "the set of elements x such that x has property p." The vertical bar ($\mid$) stands for "such that".

Suppose you wanted to write a report on an animal in a zoo that does not belong to the cat family. If we asked you to write about an object that is not in the cat family, you might write about your friend, a painting, or gold. To avoid that, we specify the overall set to which the elements of all other sets must belong. Such as set is called a **universal set**. In general, the universal set, denoted by **U**, is a set that contains all the elements that are being considered in a given discussion.

Thus, in the discussion above, if

$$U = \{x \mid x \text{ is an animal in a given zoo}\}$$

and

$$S = \{y \mid y \text{ is in the cat family}\}$$

then you are asked to write about an element in the set

$$\overline{S}=\{z\,|\,z\in U \text{ and } z\notin S\}$$

The set $\overline{S}$ is called the **complement** of S.

Note that the set S has the property that for every element $y\in S$, $y\in U$. This leads to the following definition.

If every element of a set A is an element of set B, then A is said to be a **subset** of B. In symbols, we write $A\subset B$. If A is not a subset of B we write $A\not\subset B$.

EXAMPLE 1. Let $U=\{a,b,c,d,e\}$ and $A=\{b,d,e\}$. Find

(a) $\overline{A}$ (b)   all subsets of $\overline{A}$

Solution.
   (a) $\overline{A}=\{a,c\}$
   (b) $\{a\}$, $\{c\}$, $\{a,c\}$, $\varnothing$.

It is by convention that the empty set is regarded as a subset of every set. The set $\{a,c\}$ is called an **improper subset** of $\overline{A}$. The rest of the subsets are called **proper subsets**.

If A and B are two sets such that $A\subset B$ and $B\subset A$, then A and B are said to be equal and we write $A=B$.

EXAMPLE 2. $\{2,4,6\}=\{6,2,4\}$.

EXAMPLE 3. $\{4,6,8\}=\{2+2,4+2,3+5\}$

EXAMPLE 4. $\{4,6,9\}\neq\{4,6\}$.

If A and B are two sets, then the set C consisting of all elements that belong to both A and B is called the **intersection** of A and B. In symbols, we write, $A\cap B=C$.

EXAMPLE 5. Let $A=\{\text{red, blue, purple}\}$, and $B=\{\text{white, red, blue, brown}\}$. Find $A\cap B$.

Solution. $A\cap B=\{\text{red, blue}\}$.

EXAMPLE 6. Let $A=\{1,3,5\}$, and $B=\{2,4,6\}$. Find $A\cap B$.

Solution. $A\cap B=\varnothing$. This is the case, because there is no elements in A that is also an element in B. Hence, the intersection of these two sets is the empty set.

Another operation involving two sets is the **union**. If A and B are sets, then the union of A and B, denoted by A∪B, is a set consisting of elements that belong to either A, or B, or both.

EXAMPLE 7. Let $A=\{1,2,3,4,5\}$ and $B=\{2,4,8\}$. Find A∪B.

Solution. $A∪B=\{1,2,3,4,5,8\}$.

EXERCISES A.1

1. Which of the following collections of objects are sets? Explain.

    (a) The tall students at your college.

    (b) The collection of all students enrolled in a mathematics course at your college.

    (c) The known planets of our solar system.

    (d) The number zero.

2. Which of the following statements are true?

    (a) $3\in\{1,2,3,4\}$ (b) $2\subset\{1,2,3,4\}$

    (c) $\{2\}\in\{1,2,3\}$ (d) $\{2,4\}=\{4,2\}$

    (e) $\varnothing\not\subset\{a,b,c\}$ (f) $\varnothing=\{0\}$

3. List the elements of each set.

    (a) $\{x\,|\,x$ is a natural number less than 11$\}$.

    (b) $\{y\,|\,y$ is an ocean on this earth$\}$.

    (c) $\{c\,|\,c$ is an odd number between 4 and 12$\}$.

    (d) $\{p\,|\,p$ was a president of the United States in 1979$\}$.

    (e) $\{x\,|\,x$ is a letter of the word *mathematics*$\}$.

4. Which of the following is true for sets A,B, and C?

$A = \{z \mid z$ is a letter of the word *blow*$\}$

$B = \{z \mid z$ is a letter of the word *bowl*$\}$

$C = \{z \mid z$ is a letter of the word *bellow*$\}$

(a) $A \subset C$ (b) $A = B$

(c) $B = C$ (d) $C \subset B$

5. List all subsets of $A = \{0,1,2,3\}$.

6. Determine a value for x so that each statement will be true.

(a) $\{2x,1\} = \{4,1\}$

(b) $\{3x\} = \{0\}$

(c) $\{x^2,1\} = \{4,1\}$

(d) $\{x,3x\} = \{3,1\}$

In Exercises 7-21, let $U = \{0,1,2,3,4,5,6,7,8,9\}$, $A = \{1,3,5\}$, $B = \{1,2,3,4,5\}$, and $C = \{4,6,8\}$. Determine each of the following:

7. $A \cup C$ 8. $A \cap C$ 9. $A \cap B$

10. $A \cap \varnothing$ 11. $\bar{B}$ 12. $(A \cup B) \cap C$

13. $\bar{A} \cup C$ 14. $\bar{C} \cap \bar{B}$ 15. $(A \cap B) \cap \bar{C}$

16. $\bar{U}$ 17. $\bar{\varnothing}$ 18. $A \cup \bar{U}$

19. $\overline{(A \cap B)}$ 20. $\overline{(A \cap \bar{B})}$ 21. $\overline{(\bar{A} \cap \bar{B})}$

22. Let $A = \{a,b,\{a,b\},c\}$ and $B = \{a,b,\{a,b\},\{a,b,c\}\}$. Which of the following are true?

(a) $\{a\} c (A \cap B)$

(b) $\{a,b\} \in (A \cap B)$

(c) $\{a,b\} \subset (A \cup B)$

(d) $\{a,c\} \subset A$

(e) $\{a,b,c\} \subset (A \cup B)$

(f) $\{a,b,c\} \subset (A \cap B)$

(g) $\{a\{a,b\}\} = A \cap B$

(h) $c \in (A \cap B)$

23. Let $U = \{1,a,b,c,d\}$. If $A \cap B = \{a,c\}$, $A \cup B = \{a,b,c,d\}$, $A \cap C = \{a,b\}$ and $A \cup D = \{1,a,b,c\}$ find A,B,C, and D. Is the set C unique? Is the set D unique?

A.2 THE LOGARITHMIC FUNCTION

OBJECTIVES

1. Define and graph exponential functions.
2. Define and graph logarithmic functions.
3. Define common and natural logarithms.
4. Use tables to find the logarithm of a number.
5. Use logarithms in numerical calculations.

The function f defined by $f(x)=b^x$ where $b>0$ and $b\neq 1$, is called an **exponential function**. The number b is called the **base** of the function. The domain of f is the set of all real numbers and the range of f is the set of all positive numbers.

To illustrate the behavior of the exponential function, we sketch the graphs for the bases 2 and $\frac{1}{3}$ respectively.

EXAMPLE 1. Sketch the graph of each of the following exponential functions.

(a) $f(x)=2^x$ (b) $f(x)=(\frac{1}{3})^x$

Solution. (a) By assigning values to x and computing the corresponding values for f(x), we obtain some points belonging to f as shown in Table 1.

x	−3	−2	−1	0	1	2	3
f(x)	⅛	¼	½	1	2	4	8

Table 1

Using Table 1, we locate the points as shown in Figure 1 and sketch a graph of the function. We are assuming that the function is continuous, which means the graph has no breaks. In fact, all exponential functions are continuous over the entire real line.

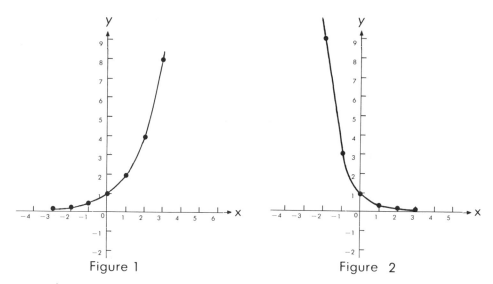

Figure 1   Figure 2

(b)   First we tabulate a few points belonging to the function $f(x)=(\frac{1}{3})^x$ in Table 2. A sketch of the graph of f is shown in Figure 2.

x	−2	−1	0	1	2	3
f(x)	9	3	1	$\frac{1}{3}$	$\frac{1}{9}$	$\frac{1}{27}$

Table 2

It should be noted that the exponential function $f(x)=b^x$ is an increasing or a decreasing function, and so, it is one-to-one. Therefore, the inverse function of f exists and has as its domain $\{x\,|\,x>0\}$ and range $\{y\,|\,y\in R\}$. The inverse of an exponential function is called a **logarithmic function**.

> **DEFINITION 1**   Suppose $b>0$ and $b\neq1$. Then the *logarithmic function* with base b is denoted by
>
> $$f^{-1}(x)=\log_b x \qquad\qquad (1)$$
>
> (read "logarithm to the base b of x"), is defined to be the inverse of the exponential function $f(x)=b^x$.

Since f^{-1} is the inverse of the function f, we have

$$x=f(f^{-1}(x))$$

Therefore, for the exponential function we obtain

$$x = b^{f^{-1}(x)} \tag{2}$$

From Equations 1 and 2, we get

$$x = b^{\log_b x} \tag{3}$$

Thus, from Equation 3, we conclude that

$$y = \log_b x \text{ if and only if } x = b^y \tag{4}$$

provided b is a positive real number and $b \neq 1$. For example:

$$\log_2 16 = 4 \text{ is equivalent to } 16 = 2^4$$
$$\log_5 1 = 0 \text{ is equivalent to } 1 = 5^0$$
$$\log_9 \tfrac{1}{3} = -\tfrac{1}{2} \text{ is equivalent to } \tfrac{1}{3} = 9^{-1/2}$$
$$10^3 = 1000 \text{ is equivalent to } \log_{10} 1000 = 3$$
$$3^{-2} = \tfrac{1}{9} \text{ is equivalent to } \log_3 \tfrac{1}{9} = -2$$

The graphs of f and f^{-1} are shown in Figure 3 for $b > 1$ and in Figure 4 for $0 < b < 1$. Observe that in each case the graphs of f^{-1} and f are symmetric with respect to the line $y = x$. Thus we obtain the graph of $y = \log_b x$ by reflecting the graph of $y = b^x$ across the line $y = x$.

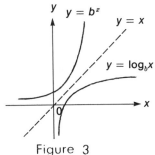

Figure 3

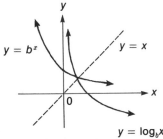

Figure 4

We now list some properties of the graph of the logarithmic function with base b.

Property 1.   For all b ($b > 0$ and $b \neq 1$), the graph of $y = \log_b x$ intersects the x-axis at $x = 1$ only.

Property 2.   If $b > 1$, then $\log_b$ is an increasing function. If $0 < b < 1$, then $\log_b$ is a decreasing function.

Property 3.   If $b > 1$, then $\log_b$ is positive for $x > 1$ and is negative for $0 < x < 1$. If $0 < b < 1$, then $\log_b$ is negative for $x > 1$ and is positive for $0 < x < 1$.

We note that the logarithmic function is one-to-one. Consequently,

$$\log_b x_1 = \log_b x_2 \text{ if and only if } x_1 = x_2$$

The following example illustrates the use of the criterion given in Equations 4.

EXAMPLE 2. Find $x, y,$ or b in each of the following equations.

(a) $y = \log_3 27$ (b) $y = \log_5(\frac{1}{5})$ (c) $4 = \log_b 81$ (d) $\frac{1}{2} = \log_9 x$

Solution.

(a)   The equation $y = \log_3 27$ is equivalent to $3^y = 27$. Hence, $3^y = 3^3$ or $y = 3$.
(b)   The equation $y = \log_5(\frac{1}{5})$ is equivalent to $5^y = \frac{1}{5}$. Hence, $5^y = 5^{-1}$ or $y = -1$.
(c)   The equation $4 = \log_b 81$ is equivalent to $b^4 = 81$. Hence, $b = \sqrt[4]{81} = 3$.
(d)   The equation $\frac{1}{2} = \log_9 x$ is equivalent to $9^{1/2} = x$. Hence, $x = 3$.

With the aid of Equation 4, the following theorem can be established readily.

THEOREM 1   If $b > 0$, $b \neq 1$, and m and n are positive numbers, then:

(a) $\log_b(mn) = \log_b m + \log_b n$.

(b) $\log_b(m/n) = \log_b m - \log_b n$.

(c) $\log 1 = 0$.

(d) $\log_b b = 1$.

(e) $\log_b m^r = r \log_b m$, where r is a real number.

EXAMPLE 3. Given $\log_{10} 2 = 0.301$ and $\log_{10} 3 = 0.477$, use the properties of logarithms in Theorem 1 to compute the value of each of the following logarithms.

(a) $\log_{10} 6$ (b) $\log_{10} 300$ (c) $\log_{10} 1.5$ (d) $\log_{10}\sqrt{0.6}$

Solution.

(a) $\log_{10} 6 = \log_{10} 2 \times 3$
$= \log_{10} 2 + \log_{10} 3$
$= 0.301 + 0.477 = 0.778$

(b) $\log_{10} 300$ **(b)** $= \log_{10} 3 \times 10^2$

$= \log_{10} 3 + \log_{10} 10^2$

$= 0.477 + 2 \log_{10} 10$

$= 0.477 + 2 = 2.477$

(c) $\log_{10} 1.5 = \log_{10}(\frac{3}{2})$

$= \log_{10} 3 - l0g_{10} 2$

$= 0.477 - 0.301 = 0.176$

(d) $\log_{10}\sqrt{0.6} = \log_{10}[(2 \times 3)/(10)]^{1/2}$

$= \frac{1}{2}[\log_{10}(2 \times 3) - \log_{10} 10]$

$= \frac{1}{2}(\log_{10} 2 + \log_{10} 3 - 1)$

$= \frac{1}{2}(0.301 + 0.477 - 1) = \frac{1}{2}(-0.222) = -0.111$

Logarithms to the base 10 and the base e are used so frequently that logarithms to these bases are abbreviated. The most common abbreviations are

$$\log_{10} x = \log x$$

and
$$\log_e x = \ln x$$

These are called the **common logarithm** and the **natural logarithm** respectively. The latter is frequently used in applications.

Historically, one of the applications of logarithms has been as an aid to numerical calculations. Using the properties of logarithms, many numerical calculations can be greatly simplified. Because our number system is base 10, the most convenient logarithm to use for numerical calculations is the common logarithm.

Using Theorem 1, we find

$$\log 10 = 1$$
$$\log 100 = \log 10^2 = 2$$
$$\log 1000 = \log 10^3 = 3$$

and so on. We also have

$$\log 1 = 0$$
$$\log 0.1 = \log 10^{-1} = -1$$
$$\log 0.01 = \log 10^{-2} = -2$$
$$\log 0.001 = \log 10^{-3} = -3$$

and so on.

To find the common logarithm of a number that is not a power of 10, special tables must be used. A complete table is included in Appendix B. In order to use the tables, we need to be able to write a number in **scientific notation**.

Any positive number N can be written in the form

$$N = a \times 10^c \qquad (5)$$

where $1 \leq a < 10$ and c is an integer. When a number is expressed as in Equation 5, it is said to be written in scientific notation. For example:

$$35 = 3.5 \times 10$$
$$16,754 = 1.6754 \times 10^4$$
$$0.053 = 5.3 \times 10^{-2}$$

From Equation 5, we have

$$\begin{aligned}
\log N &= \log(a \times 10^c) \\
&= \log a + \log 10^c \\
&= \log a + c \qquad (6)
\end{aligned}$$

It follows from Equation 6 that the logarithm of any positive number N can be written as the sum of $\log a$, where a is between 1 and 10, plus an integer c. It is convenient to give special names to the two parts in Equation 6. We call c the **characteristic** of the logarithm and $\log a (1 \leq a < 10)$ is called the **mantissa** of the logarithm.

The characteristic of the common logarithm of N is the same as the exponent of 10 when N is written in scientific notation. Clearly, the characteristic is determined by the position of the decimal point. For example:

$$\log 2640 = \log 2.64 \times 10^3 = \log 2.64 + 3$$
$$\log 0.0264 = \log 2.64 \times 10^{-2} = \log 2.64 + (-2)$$

The characteristics are 3 and -2 respectively. Since $1 \leq a < 10$ and $\log a$ is an increasing function, we have

$$\log 1 \leq \log a < \log 10$$

or

$$0 \leq \log a < 1$$

Therefore, the mantissa of the logarithm is a nonnegative number less than 1.

We can find $\log 2.64$ from Table 2 in Appendix B. To find $\log 2.64$, find 26 in the left-hand column and find 4 across the top row. This gives the four decimal entry .4216, so that $\log 2.64 = 0.4216$

The entries in the table have been computed using advanced mathematical calculations.

EXAMPLE 4. Find each of the following common logarithms:

    (a) log 283   (b) log 0.00283

Solution. We use the table above.

$$
\begin{aligned}
\text{(a)} \qquad \log 283 &= \log(2.83 \times 10^2) \\
&= \log 2.83 + \log 10^2 \\
&= 0.4518 + 2 \\
&= 2.4518
\end{aligned}
$$

$$
\begin{aligned}
\text{(b)} \quad \log 0.00283 &= \log(2.83 \times 10^{-3}) \\
&= \log 2.83 + \log 10^{-3} \\
&= 0.4518 + (-3)
\end{aligned}
$$

The characteristic in Part (b) is -3. If we add $0.4518 + (-3)$, we get -2.5482. This is not in the form of a nonnegative mantissa and an interger. Thus, if a logarithm has a negative characteristic, we either write it in standard form as shown in Part (b) above, or rewrite it so the decimal part is nonnegative.

For example, by convention Part (b) is written as

$$
\log 0.00283 = 7.4518 - 10
$$

When using logarithms in numerical computations, we will need to reverse the above process. That is, if $\log x$ is given, we may need to find x. We call x the **antilogarithm** of y *(denoted by $x =$ antilog y)* and, by definition,

$$
x = \text{antilog } y \quad \text{if and only if} \quad \log x = y
$$

For example:

    (a) If $\log x = 4$, then $x = 10^4$. This is because $\log 10^4 = 4 \log 10 = 4$.

    (b) If $\log x = -3$, then $x = 0.001$. This is because $\log 0.001 = \log 10^{-3} = -3$.

However, if $\log x = 3.7803$, then it is not so easy to find x. The following example illustrates how the antilogarithm is found in general.

EXAMPLE 5. Find each of the following logarithms.

    (a) antilog 3.7803   (b) antilog(6.6542 − 10)   (c) antilog(−2.2668)

Solution.

(a) Let $x=$ antilog 3.7803. This means that log $x=3.7803$. The mantissa 0.7803 determines the sequence of the digits in x and the characteristic 3 determines the location of the decimal point. From the main body of Table 2, we find that the number 0.7803 is associated with 6.03. That is, log $6.03=0.7803$.

$$\log x = 3.7803$$
$$= 3+0.7803$$
$$= \log 10^3 + \log 6.03$$
$$= \log(6.03 \times 10^3)$$
$$= \log 6030$$

Therefore, $x=6{,}030$

(b) Let $x=$ antilog$(6.6542-10)$.

$$\log x = 6.6542-10$$
$$= 0.6542-4$$
$$= \log 4.51 + \log 10^{-4}$$
$$= \log(4.51 \times 10^{-4})$$
$$= \log 0.000451$$

Therefore, $x= 0.000451$.

(c) Let $x=-2.2668$.

$$\log x = -2.2668$$
$$= -2.2668+3-3$$
$$= 0.7332-3$$
$$= \log 5.41 + \log 10^{-3}$$
$$= \log(5.41 \times 10^{-3})$$
$$= \log 0.00541$$

Therefore, $x = 0.00541$

Suppose we wish to find log 2.357. Table 2 does not give a value for log 2.357. However, it does give a value for log 2.35 and log 2.36. Since $2.35 < 2.357 < 2.36$, then

$$\log 2.35 < \log 2.357 < \log 2.36$$

Thus, we may be able to approximate log 2.357. One method of approximating the logarithm of a number having four significant digits from Table 2 is called **linear interpolation**. We illustrate this method in the following example.

EXAMPLE 6. Find log 1.347.

Solution. Using Table 2, we find log 1.34 and log 1.35 and set up the following table.

$$0.01\left[0.007\begin{bmatrix}\begin{array}{c|c} x & \log x \\ \hline 1.340 & 0.1271 \\ 1.347 & y \\ 1.350 & 0.1303 \end{array}\end{bmatrix}b\right]0.0032$$

The numbers on each side represent the differences between the indicated pairs of numbers. We need to find y. The difference between y and 0.1271 is approximated by b. Next, we assume that the change in x is proportionate to the change in log x. Thus we have,

$$b/0.032 = (0.007)/(0.01)$$

Solving for b, we obtain

$$b = [(0.007)/(0.01)](0.0032) \approx 0.0022$$

Hence

$$y = \log 1.347 = 0.1271 + 0.0022 = 0.1293$$

In a similar manner, linear interpolation can be used to find antilogarithms. In the past, the application of logarithms to numerical calculations was of great importance. However, with the wide-spread use of electronic calculators, the need has decreased. In spite of the fact that the calculators have cut down on the mechanical calculation, the student should still be aware of how the laws of logarithms help in deriving an answer. The following example shows how logarithms can simplify a numerical calculation.

EXAMPLE 7. Use logarithms to compute each of the following to four significant figures.

(a) $(3.48)^7$ (b) $3^{1.62}$ (c) $\sqrt[4]{27.56}$ (d) $[(2.19)^8(\sqrt{435})]/(-6.81)^3$

Solution.

(a)   Let $x = (3.48)^7$.
$$\begin{aligned} \log x &= \log(3.48)^7 \\ &= 7 \log(3.48) \\ &= 7(0.5416) \\ &= 3.7912 \\ x &= \text{antilog } 3.7912 \\ &= 6{,}183 \end{aligned}$$

(b) Let $x = 3^{1.62}$

$$\begin{aligned} \log x &= \log 3^{1.62} \\ &= 1.62 \log 3 \\ &= (1.62)(0.4771) \\ &= 0.7729 \\ x &= \text{antilog } 0.7729 \\ &= 5.928 \end{aligned}$$

(c) Let $x = \sqrt[4]{27.56}$.

$$\begin{aligned} \log x &= \log \sqrt[4]{27.56} \\ &= \tfrac{1}{4}\log 27.56 \\ &= \tfrac{1}{4}(1.4403) \\ &= 0.3601 \\ x &= \text{antilog } 0.3601 \\ &= 2.291 \end{aligned}$$

(d) The logarithm of a negative number is not defined. Therefore, we perform the calculations with positive factors, adding the correct sign to the result.

Let $x = [(2.19)^8(\sqrt{435})]/(6.81)^3$

$$\begin{aligned} \log x &= \log[(2.19)^8)\sqrt{435})]/(6.81)^3 \\ &= 8 \log 2.19 + \tfrac{1}{2}\log 435 - 3 \log 6.81 \\ &= 8(0.3404) + \tfrac{1}{2}(2.6385) - 3(0.8331) \\ &= 2.7232 + 1.3192 - 2.4993 \\ &= 1.5431 \\ x &= \text{antilog } 1.5431 \\ &= 34.93 \end{aligned}$$

Going back to the original equation, we find that $(-6.81)^3$ is negative. The numerator of the original fraction is positive, which gives a negative result. Therefore, the answer is -34.93.

EXERCISES A.2

In Exercises 1-8, express each statement by an equivalent logarithmic statement.

1. $2^4 = 16$ 2. $3^5 = 243$ 3. $5^{-2} = \frac{1}{25}$ 4. $(\frac{1}{2})^{-3} = 8$

5. $10^{-4} = 0.0001$ 6. $\sqrt{16} = 4$ 7. $32^{3/5} = 8$ 8. $5^0 = 1$

In Exercises 9-14, express each logarithmic statement by an equivalent exponential statement.

9. $\log_3 81 = 4$ 10. $\log_{10} 1000 = 3$ 11. $\log_4 \frac{1}{2} = -\frac{1}{2}$

12. $\log_{10}(1/100) = -2$ 13. $\log_{1/3} 9 = -2$ 14. $\log_{16} \frac{1}{8} = -\frac{3}{4}$

In Exercises 15-25, find the unknown $x, y,$ or b.

15. $y = \log_8 \frac{1}{2}$ 16. $\frac{1}{2} = \log_4 x$ 17. $-4 = \log_b 16$

18. $y = \ln \sqrt[3]{e}$ 19. $y = \log 10,000$ 20. $-\frac{3}{2} = \log_4 x$

21. $\log(2x+1) = 1$ 22. $\log_2(x+1) - \log_2(x-2) = 3$

23. $\log(x+15) + \log x = 2$ 24. $\log_3(2x+1) - \log_3(x-1) = 2$

25. $\log_3 3 + \log_3 x + \log_3(2x-3) - 4 = 0$

26. Prove Part b of Theorem 1.

27. Prove Part c of Theorem 1.

28. Prove Part d of Theorem 1.

29. Show that $x^x = b^{x \, \log_b x}$.

30. (a) Show that $\log_b x = (\log_a x)/(\log_a b)$.
    (b) Show that $\log_b a = 1/\log_a b$.

In Exercises 31-34, simplify each expression.

31. $\log_5(\log_7 7)$ 32. $\log_2(\log_5 125)$ 33. $\log_3(\log_3 81)$

34. $\log_b(\log_b b)$

35.   Given log 2=0.301, log 3=0.477, and log 5=0.699. Use the pro-
      perties of logarithms to evaluate each of the following:

(a)   log ⅔   (b)   (log 2)/(log 3)

(c)   log 2^3 (d)   (log 2)^3

(e)   log 15   (f)   log 30

36.   Write each number in scientific notation.

(a)   35.62   (b)   3561   (c)   0.0071

In Exercises 37-42, find each logarithm.

37.   log 426   38.   log 6.23   39.   log 15.8

40.   log 0.0032   41.   log 0.135   42.   log 248,000

In Exercises 43-50, find each antilogarithm.

43.   antilog 2.8463   44.   antilog 1.6042   45.   antilog(9.6542−10)

46.   antilog(17.8202−20)   47.   antilog(7.1847−10)

48.   antilog 5.9440   49.   antilog 0.5441   50.   antilog(−1.6108)

In Exercises 51-56, use interpolation in Table 2, to find each logarithm.

51.   log 3.654   52.   log 53.56   53.   log 0.2612

54.   log 3672   55.   log 0.0009354 56.   log 30,040

In Exercises 57-62, use interpolation in Table 2 to find x to four significant
figures.

57.   log x=2.5473   58.   log x=0.2345   59.   log x=1.7351

60.   log x=8.6543−10   61.   log x=7.8888−10

62.   log x=3.3232

In Exercises 63-78, use logarithms to perform each computaion to four
significant figures.

63. $(23.7)/(0.417)^{1/2}$ 64.   (4550)/(0.0321)   65. $1/(2.13)^4$

66. $1/(0.768)^8$ 67.   (−2.34)(3.43)(21.7)

68. $[(46,800)(5360)]/[(651)^2(−0.42)]$ 69. $(2220)^{3/2}$

70. $5^{7/3}$ 71. $\sqrt[5]{2001}$ 72. $\sqrt{(87.36)/(486.3)}$

73. $\sqrt[3]{(25.71)(0.3104)}$

74. $\sqrt{[(14.52)(66.3)/(12.26)]}$

75. $[(\sqrt{4833})(0.0123)/(471.2)]^{1/2}$

76. $\log(\log 28.4)$

77. $(\log 0.9314)/(\log 5.137)$

78. $(\log 84.6)/(\log 41.5)$

In Exercises 79-84, solve for x.

79. $x = \log_4 18$

80. $2^{2x-1} = 0.07$

81. $3^{5x} = 4$

82. $(1.03)^x = 1.782$

83. $x = \log_2 500$

84. $x = \log_6 45$

APPENDIX B:
TABLES

VALUES OF TRIGONOMETRIC FUNCTIONS

Angle θ									
Degrees	Radians	sin θ	csc θ	tan θ	cot θ	sec θ	cos θ	Radians	Degrees
0°00′	.0000	.0000	No Value	.0000	No Value	1.000	1.0000	1.5708	90°00′
10	.0029	.0029	343.8	.0029	343.8	1.000	1.0000	1.5679	50
20	.0058	.0058	171.9	.0058	171.9	1.000	1.0000	1.5650	40
30	.0087	.0087	114.6	.0087	114.6	1.000	1.0000	1.5621	30
40	.0116	.0116	85.95	.0116	85.94	1.000	.9999	1.5592	20
50	.0145	.0145	68.76	.0145	68.75	1.000	.9999	1.5563	10
1°00′	.0175	.0175	57.30	.0175	57.39	1.000	.9998	1.5533	89°00′
10	.0204	.0204	49.11	.0204	49.10	1.000	.9998	1.5504	50
20	.0233	.0233	42.98	.0233	42.96	1.000	.9997	1.5475	40
30	.0262	.0262	38.20	.0262	38.19	1.000	.9997	1.5446	30
40	.0291	.0291	34.38	.0291	34.37	1.000	.9996	1.5417	20
50	.0320	.0320	31.26	.0320	31.24	1.001	.9995	1.5388	10
2°00′	.0349	.0349	28.65	.0349	28.64	1.001	.9994	1.5359	88°00′
10	.0378	.0378	26.45	.0378	26.43	1.001	.9993	1.5330	50
20	.0407	.0407	24.56	.0407	24.54	1.001	.9992	1.5301	40
30	.0436	.0436	22.93	.0437	22.90	1.001	.9990	1.5272	30
40	.0465	.0465	21.49	.0466	21.47	1.001	.9989	1.5243	20
50	.0495	.0494	20.23	.0495	20.21	1.001	.9988	1.5213	10
3°00′	.0524	.0523	19.11	.0524	19.08	1.001	.9986	1.5184	87°00′
10	.0553	.0552	18.10	.0553	18.07	1.002	.9985	1.5155	50
20	.0582	.0581	17.20	.0582	17.17	1.002	.9983	1.5126	40
30	.0611	.0610	16.38	.0612	16.35	1.002	.9981	1.5097	30
40	.0640	.0640	15.64	.0641	15.60	1.002	.9980	1.5068	20
50	.0669	.0669	14.96	.0670	14.92	1.002	.9978	1.5039	10
4°00′	.0698	.0698	14.34	.0699	14.30	1.002	.9976	1.5010	86°00′
10	.0727	.0727	13.76	.0729	13.73	1.003	.9974	1.4981	50
20	.0756	.0765	13.23	.0758	13.20	1.003	.9971	1.4952	40
30	.0785	.0785	12.75	.0787	12.71	1.003	.9969	1.4923	30
40	.0814	.0814	12.29	.0816	12.25	1.003	.9967	1.4893	20
50	.0844	.0843	11.87	.0846	11.83	1.004	.9964	1.4864	10
Degrees	Radians	cos θ	sec θ	cot θ	tan θ	csc θ	sin θ	Radians	Degrees
								Angle θ	

Angle θ		sin θ	csc θ	tan θ	cot θ	sec θ	cos θ	Radians	Degrees
Degrees	Radians								
5°00′	.0873	.0872	11.47	.0875	11.43	1.004	.9962	1.4835	85°00′
10	.0902	.0901	11.10	.0904	11.06	1.004	.9959	1.4806	50
20	.0931	.0929	10.76	.0934	10.71	1.004	.9957	1.4777	40
30	.0960	.0958	10.43	.0963	10.39	1.005	.9954	1.4748	30
40	.0989	.0987	10.13	.0992	10.08	1.005	.9951	1.4719	20
50	.1018	.1016	9.839	.1022	9.788	1.005	.9948	1.4690	10
6°00′	.1047	.1045	9.567	.1051	9.514	1.006	.9945	1.4661	84°00′
10	.1076	.1074	9.309	.1080	9.255	1.006	.9942	1.4632	50
20	.1105	.1103	9.065	.1110	9.010	1.006	.9939	1.4603	40
30	.1134	.1132	8.834	.1139	8.777	1.006	.9936	1.4573	30
40	.1164	.1161	8.614	.1169	8.556	1.007	.9932	1.4544	20
50	.1193	.1190	8.405	.1198	8.345	1.007	.9929	1.4515	10
7°00′	.1222	.1219	8.206	.1228	8.144	1.008	.9925	1.4486	83°00′
10	.1251	.1248	8.016	.1257	7.953	1.008	.9922	1.4457	50
20	.1280	.1276	7.834	.1287	7.770	1.008	.9918	1.4428	40
30	.1309	.1305	7.661	.1317	7.596	1.009	.9914	1.4399	30
40	.1338	.1334	7.496	.1346	7.429	1.009	.9911	1.4370	20
50	.1367	.1363	7.337	.1376	7.269	1.009	.9907	1.4341	10
8°00′	.1396	.1392	7.185	.1405	7.115	1.010	.9903	1.4312	82°00′
10	.1425	.1421	7.040	.1435	6.968	1.010	.9899	1.4283	50
20	.1454	.1449	6.900	.1465	6.827	1.011	.9894	1.4254	40
30	.1484	.1478	6.765	.1495	6.691	1.011	.9890	1.4224	30
40	.1513	.1507	6.636	.1524	6.561	1.012	.9886	1.4195	20
50	.1542	.1536	6.512	.1554	6.435	1.012	.9881	1.4166	10
9°00′	.1571	.1564	6.392	.1584	6.314	1.012	.9877	1.4137	81°00′
10	.1600	.1593	6.277	.1614	6.197	1.013	.9872	1.4108	50
20	.1629	.1622	6.166	.1644	6.084	1.013	.9868	1.4079	40
30	.1658	.1650	6.059	.1673	5.976	1.014	.9863	1.4050	30
40	.1687	.1679	5.955	.1703	5.871	1.014	.9858	1.4021	20
50	.1716	.1708	5.855	.1733	5.769	1.015	.9853	1.3992	10
10°00′	.1745	.1736	5.759	.1763	5.671	1.015	.9848	1.3963	80°00′
10	.1774	.1765	5.665	.1793	5.576	1.016	.9843	1.3934	50
20	.1804	.1794	5.575	.1823	5.485	1.016	.9838	1.3904	40
30	.1833	.1822	5.487	.1853	5.396	1.017	.9833	1.3875	30
40	.1862	.1851	5.403	.1883	5.309	1.018	.9827	1.3846	20
50	.1891	.1880	5.320	.1914	5.226	1.018	.9822	1.3817	10
11°00′	.1920	.1908	5.241	.1944	5.145	1.019	.9816	1.3788	79°00′
10	.1949	.1937	5.164	.1974	5.066	1.019	.9811	1.3759	50
20	.1978	.1965	5.089	.2004	4.989	1.020	.9805	1.3730	40
30	.2007	.1994	5.016	.2035	4.915	1.020	.9799	1.3701	30
40	.2036	.2022	4.945	.2065	4.843	1.021	.9793	1.3672	20
50	.2065	.2051	4.876	.2095	4.773	1.022	.9787	1.3643	10
12°00′	.2094	.2079	4.810	.2126	4.705	1.022	.9781	1.3614	78°00′
10	.2123	.2108	4.745	.2156	4.638	1.023	.9775	1.3584	50
20	.2153	.2136	4.682	.2186	4.574	1.024	.9769	1.3555	40
30	.2182	.2164	4.620	.2217	4.511	1.024	.9763	1.3526	30
40	.2211	.2193	4.560	.2247	4.449	1.025	.9757	1.3497	20
50	.2240	.2221	4.502	.2278	4.390	1.026	.9750	1.3468	10
Degrees	Radians	cos θ	sec θ	cot θ	tan θ	csc θ	sin θ	Radians	Degrees
								Angle θ	

Angle θ									
Degrees	Radians	sin θ	csc θ	tan θ	cot θ	sec θ	cos θ	Radians	Degrees
13°00′	.2269	.2250	4.445	.2309	4.331	1.026	.9744	1.3439	77°00′
10	.2298	.2278	4.390	.2339	4.275	1.027	.9737	1.3410	50
20	.2327	.2306	4.336	.2370	4.219	1.028	.9730	1.3381	40
30	.2356	.2334	4.284	.2401	4.165	1.028	.9724	1.3352	30
40	.2385	.2363	4.232	.2432	4.113	1.029	.9717	1.3323	20
50	.2414	.2391	4.182	.2462	4.061	1.030	.9710	1.3294	10
14°00′	.2443	.2419	4.134	.2493	4.011	1.031	.9703	1.3265	76°00′
10	.2473	.2447	4.086	.2524	3.962	1.031	.9696	1.3235	50
20	.2502	.2476	4.039	.2555	3.914	1.032	.9689	1.3206	40
30	.2531	.2504	3.994	.2586	3.867	1.033	.9681	1.3177	30
40	.2560	.2532	3.950	.2617	3.821	1.034	.9674	1.3148	20
50	.2589	.2560	3.906	.2648	3.776	1.034	.9667	1.3119	10
15°00′	.2618	.2588	3.864	.2679	3.732	1.035	.9659	1.3090	75°00′
10	.2647	.2616	3.822	.2711	3.689	1.036	.9652	1.3061	50
20	.2676	.2644	3.782	.2742	3.647	1.037	.9644	1.3032	40
30	.2705	.2672	3.742	.2773	3.606	1.038	.9636	1.3003	30
40	.2734	.2700	3.703	.2805	3.566	1.039	.9628	1.2974	20
50	.2763	.2728	3.665	.2836	3.526	1.039	.9621	1.2945	10
16°00′	.2793	.2756	3.628	.2867	3.487	1.040	.9613	1.2915	74°00′
10	.2822	.2784	3.592	2.899	3.450	1.041	.9605	1.2886	50
20	.2851	.2812	3.556	.2931	3.412	1.042	.9596	1.2857	40
30	.2880	.2840	3.521	.2962	3.376	1.043	.9588	1.2828	30
40	.2909	.2868	3.487	.2944	3.340	1.044	.9580	1.2799	20
50	.2938	.2896	3.453	.3026	3.305	1.045	.9572	1.2770	10
17°00′	.2967	.2924	3.420	.3057	3.271	1.046	.9563	1.2741	73°00′
10	.2996	.2952	3.388	.3089	3.237	1.047	.9555	1.2712	50
20	.3025	.2979	3.357	.3121	3.204	1.048	.9546	1.2683	40
30	.3054	.3007	3.326	.3153	3.172	1.048	.9537	1.2654	30
40	.3083	.3035	3.295	.3185	3.140	1.049	.9528	1.2625	20
50	.3113	.3062	3.265	.3217	3.108	1.050	.9520	1.2595	10
18°00′	.3142	.3090	3.236	.3249	3.078	1.051	.9511	1.2566	72°00′
10	.3171	.3118	3.207	.3281	3.047	1.052	.9502	1.2537	50
20	.3200	.3145	3.179	.3314	3.018	1.053	.9492	1.2508	40
30	.3229	.3173	3.152	.3346	2.989	1.054	.9483	1.2479	30
40	.3258	.3201	3.124	.3378	2.960	1.056	.9474	1.2450	20
50	.3287	.3228	3.098	.3411	2.932	1.057	.9465	1.2421	10
19°00′	.3316	.3256	3.072	.3443	2.904	1.058	.9455	1.2392	71°00
10	.3345	.3283	3.046	.3476	2.877	1.059	.9446	1.2363	50
20	.3374	.3311	3.021	.3508	2.850	1.060	.9436	1.2334	40
30	.3403	.3338	2.996	.3541	2.824	1.061	.9426	1.2305	30
40	.3432	.3365	2.971	.3574	2.798	1.062	.9417	1.2275	20
50	.3462	.3393	2.947	.3607	2.773	1.063	.9407	1.2246	10
20°00′	.3491	.3420	2.924	.3640	2.747	1.064	.9397	1.2217	70°00′
10	.3520	.3448	2.901	.3673	2.723	1.065	.9387	1.2188	50
20	.3549	.3475	2.878	.3706	2.699	1.066	.9377	1.2159	40
30	.3578	.3502	2.855	.3739	2.675	1.068	.9367	1.2130	30
40	.3607	.3529	2.833	.3772	2.651	1.069	.9356	1.2101	20
50	.3636	.3557	2.812	.3805	2.628	1.070	.9346	1.2072	10
Degrees	Radians	cos θ	sec θ	cot θ	tan θ	csc θ	sin θ	Radians	Degrees

| | | | | | | | | Angle θ | |

Angle θ		sin θ	csc θ	tan θ	cot θ	sec θ	cos θ	Radians	Degrees
Degrees	Radians								
21°00′	.3665	.3584	2.790	.3839	2.605	1.071	.9336	1.2043	69°00′
10	.3694	.3611	2.769	.3872	2.583	1.072	.9325	1.2014	50
20	.3723	.3638	2.749	.3906	2.560	1.074	.9315	1.1985	40
30	.3752	.3665	2.729	.3939	2.539	1.075	.9304	1.1956	30
40	.3782	.3692	2.709	.3973	2.517	1.076	.9293	1.1926	20
50	.3811	.3719	2.689	.4006	2.496	1.077	.9283	1.1897	10
22°00′	.3840	.3746	2.669	.4040	2.475	1.079	.9272	1.1868	68°00′
10	.3869	.3773	2.650	.4074	2.455	1.080	.9261	1.1839	50
20	.3898	.3880	2.632	.4108	2.434	1.081	.9250	1.1810	40
30	.3927	.3827	2.613	.4142	2.414	1.082	.9239	1.1781	30
40	.3956	.3854	2.595	.4176	2.394	1.084	.9228	1.1752	20
50	.3985	.3881	2.557	.4210	2.375	1.085	.9216	1.1723	10
23°00′	.4014	.3907	2.559	.4245	2.356	1.086	.9205	1.1694	67°00′
10	.4043	.3934	2.542	.4279	2.337	1.088	.9194	1.1665	50
20	.4072	.3961	2.525	.4314	2.318	1.089	.9182	1.1636	40
30	.4102	.3987	2.508	.4348	2.300	1.090	.9171	1.1606	30
40	.4131	.4014	2.491	.4383	2.282	1.092	.9159	1.1577	20
50	.4160	.4041	2.475	.4417	2.264	1.093	.9147	1.1548	10
24°00′	.4189	.4067	2.459	.4452	2.246	1.095	.9135	1.1519	66°00′
10	.4218	.4094	2.443	.4487	2.229	1.096	.9124	1.1490	50
20	.4247	.4120	2.427	.4522	2.211	1.097	.9112	1.1461	40
30	.4276	.4147	2.411	.4557	2.194	1.099	.9100	1.1432	30
40	.4305	.4173	2.396	.4592	2.177	1.100	.9088	1.1403	20
50	.4334	.4200	2.381	.4628	2.161	1.102	.9075	1.2374	10
25°00′	.4363	.4226	2.366	.4663	2.145	1.103	.9063	1.1345	65°00′
10	.4392	.4253	2.352	.4699	2.128	1.105	.9051	1.1316	50
20	.4422	.4279	2.337	.4734	2.112	1.106	.9038	1.1286	40
30	.4451	.4305	2.323	.4770	2.097	1.108	.9026	1.1257	30
40	.4480	.4331	2.309	.4806	2.081	1.109	.9013	1.1228	20
50	.4509	.4358	2.295	.4841	2.066	1.111	.9001	1.1199	10
26°00′	.4538	.4384	2.281	.4877	2.050	1.113	.8988	1.1170	64°00′
10	.4567	.4410	2.268	.4913	2.035	1.114	.8975	1.1141	50
20	.4596	.4436	2.254	.4950	2.020	1.116	.8962	1.1112	40
30	.4625	.4462	2.241	.4986	2.006	1.117	.8949	1.1083	30
40	.4654	.4488	2.228	.5022	1.991	1.119	.8936	1.1054	20
50	.4683	.4514	2.215	.5059	1.977	1.121	.8923	1.1025	10
27°00′	.4712	.4540	2.203	.5095	1.963	1.122	.8910	1.0996	63°00′
10	.4741	.4566	2.190	.5132	1.949	1.124	.8897	1.0966	50
20	.4771	.4592	2.178	.5169	1.935	1.126	.8884	1.0937	40
30	.4800	.4617	2.166	.5206	1.921	1.127	.8870	1.0908	30
40	.4829	.4643	2.154	.5243	1.907	1.129	.8857	1.0879	20
50	.4858	.4669	2.142	.5280	1.894	1.131	.8843	1.0850	10
28°00′	.4887	.4695	2.130	.5317	1.881	1.133	.8829	1.0821	62°00′
10	.4916	.4720	2.118	.5354	1.868	1.134	.8816	1.0792	50
20	.4945	.4746	2.107	.5392	1.855	1.136	.8802	1.0763	40
30	.4947	.4772	2.096	.5430	1.842	1.138	.8788	1.0734	30
40	.5003	.4797	2.085	.5467	1.829	1.140	.8774	1.0705	20
50	.5032	.4823	2.074	.5505	1.816	1.142	.8760	1.0676	10
Degrees	Radians	cos θ	sec θ	cot θ	tan θ	csc θ	sin θ	Radians	Degrees

		Angle θ

TABLE ONE

Angle θ									
Degrees	Radians	cos θ	sec θ	cot θ	tan θ	csc θ	sin θ	Radians	Degrees
29°00′	.5061	.4848	2.063	.5543	1.804	1.143	.8746	1.0647	61°00′
10	.5091	.4874	2.052	.5581	1.792	1.145	.8732	1.0617	50
20	.5120	.4899	2.041	.5619	1.780	1.147	.8718	1.0588	40
30	.5149	.4924	2.031	.5658	1.767	1.149	.8704	1.0559	30
40	.5178	.4950	2.020	.5696	1.756	1.151	.8689	1.0530	20
50	.5207	.4975	2.010	.5735	1.744	1.153	.8675	1.0501	10
30°00′	.5236	.5000	2.000	.5774	1.732	1.155	.8660	1.0472	60°00′
10	.5265	.5025	1.990	.5812	1.720	1.157	.8646	1.0443	50
20	.5294	.5050	1.980	.5851	1.709	1.159	.8631	1.0414	40
30	.5323	.5075	1.970	.5890	1.698	1.161	.8616	1.0385	30
40	.5352	.5100	1.961	.5930	1.686	1.163	.8601	1.0356	20
50	.5381	.5125	1.951	.5969	1.675	1.165	.8587	1.0327	10
31°00′	.5411	.5150	1.942	.6009	1.664	1.167	.8572	1.0297	59°00′
10	.5440	.5175	1.932	.6048	1.653	1.169	.8557	1.0268	50
20	.5469	.5200	1.923	.6088	1.643	1.171	.8542	1.0239	40
30	.5498	.5225	1.914	.6128	1.632	1.173	.8526	1.0210	30
40	.5527	.5250	1.905	.6168	1.621	1.175	.8511	1.0181	20
50	.5556	.5275	1.896	.6208	1.611	1.177	.8496	1.0152	10
32°00′	.5585	.5299	1.887	.6249	1.600	1.179	.8480	1.0123	58°00′
10	.5614	.5324	1.878	.6289	1.590	1.181	.8465	1.0094	50
20	.5643	.5348	1.870	.6330	1.580	1.184	.8450	1.0065	40
30	.5672	.5373	1.861	.6371	1.570	1.186	.8434	1.0036	30
40	.5701	.5398	1.853	.6412	1.560	1.188	.8418	1.0007	20
50	.5730	.5422	1.844	.6452	1.550	1.190	.8403	.9977	10
33°00′	.5760	.5446	1.836	.6494	1.540	1.192	.8387	.9948	57°00′
10	.5789	.5471	1.828	.6536	1.530	1.195	.8371	.9919	50
20	.5818	.5495	1.820	.6577	1.520	1.197	.8355	.9890	40
30	.5847	.5519	1.812	.6619	1.511	1.199	.8339	.9861	30
40	.5876	.5544	1.804	.6661	1.501	1.202	.8323	.9832	20
50	.5905	.5568	1.796	.6703	1.492	1.204	.8307	.9803	10
34°00′	.5934	.5592	1.788	.6745	1.483	1.206	.8290	.9774	56°00′
10	.5963	.5616	1.781	.6787	1.473	1.209	.8274	.9745	50
20	.5992	.5640	1.773	.6830	1.464	1.211	.8258	.9716	40
30	.6021	.5664	1.766	.6873	1.455	1.213	.8241	.9687	30
40	.6050	.5688	1.758	.6916	1.446	1.216	.8225	.9657	20
50	.6080	.5712	1.751	.6959	1.437	1.218	.8208	.9628	10
35°00′	.6109	.5736	1.743	.7002	1.428	1.221	.8192	.9599	55°00′
10	.6138	.5760	1.736	.7046	1.419	1.223	.8175	.9570	50
20	.6167	.5783	1.729	.7089	1.411	1.226	.8158	.9541	40
30	.6196	.5807	1.722	.7133	1.402	1.228	.8141	.9512	30
40	.6225	.5831	1.715	.7177	1.393	1.231	.8124	.9483	20
50	.6254	.5854	1.708	.7221	1.385	1.233	.8107	.9454	10
36°00′	.6283	.5878	1.701	.7265	1.376	1.236	.8090	.9425	54°00′
10	.6312	.5901	1.695	.7310	1.368	1.239	.8073	.9396	50
20	.6341	.5925	1.688	.7355	1.360	1.241	.8056	.9367	40
30	.6370	.5948	1.681	.7400	1.351	1.244	.8039	.9338	30
40	.6400	.5972	1.675	.7445	1.343	1.247	.8021	.9308	20
50	.6429	.5995	1.668	.7490	1.335	1.249	.8004	.9279	10
Degrees	Radians	sin θ	csc θ	tan θ	cot θ	sec θ	cos θ	Radians	Degrees

Angle θ

Angle θ		sin θ	csc θ	tan θ	cot θ	sec θ	cos θ		
Degrees	Radians							Radians	Degrees
37°00′	.6458	.6018	1.662	.7536	1.327	1.252	.7986	.9250	53°00′
10	.6487	.6041	1.655	.7581	1.319	1.255	.7696	.9221	50
20	.6516	.6065	1.649	.7627	1.311	1.258	.7951	.9192	40
30	.6545	.6088	1.643	.7673	1.303	1.260	.7934	.9163	30
40	.6574	.6111	1.636	.7720	1.295	1.263	.7916	.9134	20
50	.6603	.6134	1.630	.7766	1.288	1.266	.7898	.9105	10
38°00′	.6632	.6157	1.624	.7813	1.280	1.269	.7880	.9076	52°00′
10	.6661	.6180	1.618	.7860	1.272	1.272	.7862	.9047	50
20	.6690	.6202	1.612	.7907	1.265	1.275	.7844	.9018	40
30	.6720	.6225	1.606	.7954	1.257	1.278	.7826	.8988	30
40	.6749	.6248	1.601	.8002	1.250	1.281	.7808	.8959	20
50	.6778	.6271	1.595	.8050	1.242	1.284	.7790	.8930	10
39°00′	.6807	.6293	1.589	.8098	1.235	1.287	.7771	.8901	51°00′
10	.6836	.6316	1.583	.8146	1.228	1.290	.7753	.8872	50
20	.6865	.6338	1.578	.8195	1.220	1.293	.7735	.8843	40
30	.6894	.6361	1.572	.8243	1.213	1.296	.7716	.8814	30
40	.6923	.6383	1.567	.8292	1.206	1.299	.7698	.8785	20
50	.6952	.6406	1.561	.8342	1.199	1.302	.7679	.8756	10
40°00′	.6981	.6428	1.556	.8391	1.192	1.305	.7660	.8727	50°00′
10	.7010	.6450	1.550	.8441	1.185	1.309	.7642	.8698	50
20	.7039	.6472	1.545	.8491	1.178	1.312	.7623	.8668	40
30	.7069	.6494	1.540	.8541	1.171	1.315	.7604	.8639	30
40	.7098	.6517	1.535	.8591	1.164	1.318	.7585	.8610	20
50	.7127	.6539	1.529	.8642	1.157	1.322	.7566	.8581	10
41°00′	.7156	.6561	1.524	.8693	1.150	1.325	.7547	.8552	49°00′
10	.7185	.6583	1.519	.8744	1.144	1.328	.7528	.8523	50
20	.7214	.6604	1.514	.8796	1.137	1.332	.7509	.8494	40
30	.7243	.6626	1.509	.8847	1.130	1.335	.7490	.8465	30
40	.7272	.6648	1.504	.8899	1.124	1.339	.7470	.8436	20
50	.7301	.6670	1.499	.8952	1.117	1.342	.7451	.8407	10
42°00′	.7330	.6691	1.494	.9004	1.111	1.346	.7431	.8378	48°00′
10	.7359	.6713	1.490	.9057	1.104	1.349	.7412	.8348	50
20	.7389	.6734	1.485	.9110	1.098	1.353	.7392	.8319	40
30	.7418	.6756	1.480	.9163	1.091	1.356	.7373	.8290	30
40	.7447	.6777	1.476	.9217	1.085	1.360	.7353	.8261	20
50	.7476	.6799	1.471	.9271	1.079	1.364	.7333	.8232	10
43°00′	.7505	.6820	1.466	.9325	1.072	1.367	.7314	.8203	47°00′
10	.7534	.6841	1.462	.9380	1.066	1.371	.7294	.8174	50
20	.7563	.6862	1.457	.9435	1.060	1.375	.7274	.8145	40
30	.7592	.6884	1.453	.9490	1.054	1.379	.7254	.8116	30
40	.7621	.6905	1.448	.9545	1.048	1.382	.7234	.8087	20
50	.7650	.6926	1.444	.9601	1.042	1.386	.7214	.8058	10
44°00′	.7679	.6947	1.440	.9657	1.036	1.390	.7193	.8029	46°00′
10	.7709	.6967	1.435	.9713	1.030	1.394	.7173	.7999	50
20	.7738	.6988	1.431	.9770	1.024	1.398	.7153	.7970	40
30	.7767	.7009	1.427	.9827	1.018	1.402	.7133	.7941	30
40	.7796	.7030	1.423	.9884	1.012	1.406	.7112	.7912	20
50	.7825	.7050	1.418	.9942	1.006	1.410	.7092	.7883	10
45°00′	.7854	.7071	1.414	1.000	1.000	1.414	.7071	.7854	45°00′
Degrees	Radians	cos θ	sec θ	cot θ	tan θ	csc θ	sin θ	Radians	Degrees
								Angle θ	

COMMON LOGARITHMS

X	0	1	2	3	4	5	6	7	8	9
1.0	.0000	.0043	.0086	.0128	.0170	.0212	.0253	.0294	.0334	.0374
1.1	.0414	.0453	.0492	.0531	.0569	.0607	.0645	.0682	.0719	.0755
1.2	.0792	.0828	.0864	.0899	.0934	.0969	.1004	.1038	.1072	.1106
1.3	.1139	.1173	.1206	.1239	.1271	.1303	.1335	.1367	.1399	.1430
1.4	.1461	.1492	.1523	.1553	.1584	.1614	.1644	.1673	.1703	.1732
1.5	.1761	.1790	.1818	.1847	.1875	.1903	.1931	.1959	.1987	.2014
1.6	.2041	.2068	.2095	.2122	.2148	.2175	.2201	.2227	.2253	.2279
1.7	.2304	.2330	.2355	.2380	.2405	.2430	.2455	.2480	.2504	.2529
1.8	.2553	.2577	.2601	.2625	.2648	.2672	.2695	.2718	.2742	.2765
1.9	.2788	.2810	.2833	.2856	.2878	.2900	.2923	.2945	.2967	.2989
2.0	.3010	.3032	.3054	.3075	.3096	.3118	.3139	.3160	.3181	.3201
2.1	.3222	.3243	.3263	.3284	.3304	.3324	.3345	.3365	.3385	.3404
2.2	.3424	.3444	.3464	.3483	.3502	.3522	.3541	.3560	.3579	.3598
2.3	.3617	.3636	.3655	.3674	.3692	.3711	.3729	.3747	.3766	.3784
2.4	.3802	.3820	.3838	.3856	.3874	.3892	.3909	.3927	.3945	.3962
2.5	.3979	.3997	.4014	.4031	.4048	.4065	.4082	.4099	.4116	.4133
2.6	.4150	.4166	.4183	.4200	.4216	.4232	.4249	.4265	.4281	.4298
2.7	.4314	.4330	.4346	.4362	.4378	.4393	.4409	.4425	.4440	.4456
2.8	.4472	.4487	.4502	.4518	.4533	.4548	.4564	.4579	.4594	.4609
2.9	.4624	.4639	.4654	.4669	.4683	.4698	.4713	.4728	.4742	.4757
3.0	.4771	.4786	.4800	.4814	.4829	.4843	.4857	.4871	.4886	.4900
3.1	.4914	.4928	.4942	.4955	.4969	.4983	.4997	.5011	.5024	.5038
3.2	.5051	.5065	.5079	.5092	.5105	.5119	.5132	.5145	.5159	.5172
3.3	.5185	.5198	.5211	.5224	.5237	.5250	.5263	.5276	.5289	.5302
3.4	.5315	.5328	.5340	.5353	.5366	.5378	.5391	.5403	.5416	.5428
3.5	.5441	.5453	.5465	.5478	.5490	.5502	.5514	.5527	.5539	.5551
3.6	.5563	.5575	.5587	.5599	.5611	.5623	.5635	.5647	.5658	.5670
3.7	.5682	.5694	.5705	.5717	.5729	.5740	.5752	.5763	.5775	.5786
3.8	.5798	.5809	.5821	.5832	.5843	.5855	.5866	.5877	.5888	.5899
3.9	.5911	.5922	.5933	.5944	.5955	.5966	.5977	.5988	.5999	.6010
4.0	.6021	.6031	.6042	.6053	.6064	.6075	.6085	.6096	.6107	.6117
4.1	.6128	.6138	.6149	.6160	.6170	.6180	.6191	.6201	.6212	.6222
4.2	.6232	.6243	.6253	.6263	.6274	.6284	.6294	.6304	.6314	.6325
4.3	.6335	.6345	.6355	.6365	.6375	.6385	.6395	.6405	.6415	.6425
4.4	.6435	.6444	.6454	.6464	.6474	.6484	.6493	.6503	.6513	.6522
4.5	.6532	.6542	.6551	.6561	.6571	.6580	.6590	.6599	.6609	.6618
4.6	.6628	.6637	.6646	.6656	.6665	.6675	.6684	.6693	.6702	.6712
4.7	.6721	.6730	.6739	.6749	.6758	.6767	.6776	.6785	.6794	.6803
4.8	.6812	.6821	.6830	.6839	.6848	.6857	.6866	.6875	.6884	.6893
4.9	.6902	.6911	.6920	.6928	.6937	.6946	.6955	.6964	.6972	.6981

X	0	1	2	3	4	5	6	7	8	9
5.0	.6990	.6998	.7007	.7016	.7024	.7033	.7042	.7050	.7059	.7067
5.1	.7076	.7084	.7093	.7101	.7110	.7118	.7126	.7135	.7143	.7152
5.2	.7160	.7168	.7177	.7185	.7193	.7202	.7210	.7218	.7226	.7235
5.3	.7243	.7251	.7259	.7267	.7275	.7284	.7292	.7300	.7308	.7316
5.4	.7324	.7332	.7340	.7348	.7356	.7364	.7372	.7380	.7388	.7396
5.5	.7404	.7412	.7419	.7427	.7435	.7443	.7451	.7459	.7466	.7474
5.6	.7482	.7490	.7497	.7505	.7513	.7520	.7528	.7536	.7543	.7551
5.7	.7559	.7566	.7574	.7582	.7589	.7597	.7604	.7612	.7619	.7627
5.8	.7634	.7642	.7649	.7657	.7664	.7672	.7679	.7686	.7694	.7701
5.9	.7709	.7716	.7723	.7731	.7738	.7745	.7752	.7760	.7767	.7774
6.0	.7782	.7789	.7796	.7803	.7810	.7818	.7825	.7832	.7839	.7846
6.1	.7853	.7860	.7868	.7875	.7882	.7889	.7896	.7903	.7910	.7917
6.2	.7924	.7931	.7938	.7945	.7952	.7959	.7966	.7973	.7980	.7987
6.3	.7993	.8000	.8007	.8014	.8021	.8028	.8035	.8041	.8048	.8055
6.4	.8062	.8069	.8075	.8082	.8089	.8096	.8102	.8109	.8116	.8122
6.5	.8129	.8136	.8142	.8149	.8156	.8162	.8169	.8176	.8182	.8189
6.6	.8195	.8202	.8209	.8215	.8222	.8228	.8235	.8241	.8248	.8254
6.7	.8261	.8267	.8274	.8280	.8287	.8293	.8299	.8306	.8312	.8319
6.8	.8225	.8231	.8338	.8244	.8251	.8257	.8263	.8270	.8376	.8382
6.9	.8388	.8395	.8401	.8407	.8414	.8420	.8426	.8432	.8439	.8445
7.0	.8451	.8457	.8463	.8470	.8476	.8482	.8488	.8494	.8500	.8506
7.1	.8513	.8519	.8525	.8531	.8537	.8543	.8549	.8555	.8561	.8567
7.2	.8573	.8579	.8585	.8591	.8597	.8603	.8609	.8615	.8621	.8627
7.3	.8633	.8639	.8645	.8651	.8657	.8663	.8669	.8675	.8681	.8686
7.4	.8692	.8698	.8704	.8710	.8716	.8722	.8727	.8733	.8739	.8745
7.5	.8751	.8756	.8762	.8768	.8774	.8779	.8785	.8791	.8797	.8802
7.6	.8808	.8814	.8820	.8825	.8831	.8837	.8842	.8848	.8854	.8859
7.7	.8865	.8871	.8876	.8882	.8887	.8893	.8899	.8904	.8910	.8915
7.8	.8921	.8927	.8932	.8938	.8943	.8949	.8954	.8960	.8965	.8971
7.9	.8976	.8982	.8987	.8993	.8998	.9004	.9009	.9015	.9020	.9025
8.0	.9031	.9036	.9042	.9047	.9053	.9058	.9063	.9069	.9074	.9079
8.1	.9085	.9090	.9096	.9101	.9106	.9112	.9117	.9122	.9128	.9133
8.2	.9138	.9143	.9149	.9154	.9159	.9165	.9170	.9175	.9180	.9186
8.3	.9191	.9196	.9201	.9206	.9212	.9217	.9222	.9227	.9232	.9238
8.4	.9243	.9248	.9253	.9258	.9263	.9269	.9274	.9279	.9284	.9289
8.5	.9294	.9299	.9304	.9309	.9315	.9320	.9325	.9330	.9335	.9340
8.6	.9345	.9350	.9355	.9360	.9365	.9370	.9375	.9380	.9385	.9390
8.7	.9395	.9400	.9405	.9410	.9415	.9420	.9425	.9430	.9435	.9440
8.8	.9445	.9450	.9455	.9460	.9465	.9469	.9474	.9479	.9484	.9489
8.9	.9494	.9499	.9504	.9509	.9513	.9518	.9523	.9528	.9533	.9538
9.0	.9542	.9547	.9552	.9557	.9562	.9566	.9571	.9576	.9581	.9586
9.1	.9590	.9595	.9600	.9605	.9609	.9614	.9619	.9624	.9628	.9633
9.2	.9638	.9643	.9647	.9652	.9657	.9661	.9666	.9671	.9675	.9680
9.3	.9685	.9689	.9694	.9699	.9703	.9708	.9713	.9717	.9722	.9727
9.4	.9731	.9736	.9741	.9745	.9750	.9754	.9759	.9763	.9768	.9773
9.5	.9777	.9782	.9786	.9791	.9795	.9800	.9805	.9809	.9814	.9818
9.6	.9823	.9827	.9832	.9836	.9841	.9845	.9850	.9854	.9859	.9863
9.7	.9868	.9872	.9877	.9881	.9886	.9890	.9894	.9899	.9903	.9908
9.8	.9912	.9917	.9921	.9926	.9930	.9934	.9939	.9943	.9948	.9952
9.9	.9956	.9961	.9965	.9969	.9974	.9978	.9983	.9987	.9991	.9996

SQUARES AND SQUARE ROOTS

x	x^2	$\sqrt{x}$	x	x^2	$\sqrt{x}$
1	1	1.000	51	2,601	7.141
2	4	1.414	52	2,704	7.211
3	9	1.732	53	2,809	7.280
4	16	2.000	54	2,916	7.348
5	25	2.236	55	3,025	7.416
6	36	2.449	56	3,136	7.483
7	49	2.646	57	3,249	7.550
8	64	2.828	58	3,364	7.616
9	81	3.000	59	3,481	7.681
10	100	3.162	60	3,600	7.746
11	121	3.317	61	3,721	7.810
12	144	3.464	62	3,844	7.874
13	169	3.606	63	3,969	7.937
14	196	3.742	64	4,096	8.000
15	225	3.873	65	4,225	8.062
16	256	4.000	66	4,356	8.124
17	289	4.123	67	4,489	8.185
18	324	4.243	68	4,624	8.246
19	361	4.359	69	4,761	8.307
20	400	4.472	70	4,900	8,367
21	441	4.583	71	5,041	8.426
22	484	4.690	72	5,184	8.485
23	529	4.796	73	5,329	8.544
24	576	4.899	74	5,476	8.602
25	625	5.000	75	5,625	8.660
26	676	5.099	76	5,776	8.718
27	729	5.196	77	5,929	8.775
28	784	5.292	78	6,084	8.832
29	841	5.385	79	6,241	8.888
30	900	5.477	80	6,400	8.944
31	961	5.568	81	6,561	9.000
32	1,024	5.657	82	6,724	9.055
33	1,089	5.745	83	6,889	9.110
34	1,156	5.831	84	7,056	9.165
35	1,225	5.916	85	7,225	9.220
36	1,296	6.000	86	7,396	9.274
37	1,369	6.083	87	7,569	9.327
38	1,444	6.164	88	7,744	9.381
39	1,521	6.245	89	7,921	9.434
40	1,600	6.325	90	8,100	9.487
41	1,681	6.403	91	8,281	9.539
42	1,764	6.481	92	8,464	9.592
43	1,849	6.557	93	8,649	9.644
44	1,936	6.633	94	8,836	9.695
45	2,025	6.708	95	9,025	9.747
46	2,116	6.782	96	9,216	9.798
47	2,209	6.856	97	9,409	9.849
48	2,304	6.928	98	9,604	9.899
49	2,401	7.000	99	9,801	9.950
50	2,500	7.071	100	10,000	10.000

METRIC - ENGLISH CONVERSIONS

To convert from	To	Multiply by
millimeters (mm)	inches (in)	0.04
centimeters (cm)	inches (in)	0.39
meters (m)	feet (ft)	3.28
meters (m)	yards (yd)	1.09
kilometers (km)	miles (mi)	0.62
grams (g)	ounces (oz)	0.035
kilograms (kg)	pounds (lb)	2.2
liters (l)	pints (pt)	2.1
liters (l)	quarts (qt)	1.06
liters (l)	gallons (gal)	0.26
inches (in)	centimeters (cm)	2.54
feet (ft)	meters (m)	0.305
yards (yd)	meters (m)	0.914
miles (mi)	kilometers (km)	1.609
ounces (oz)	grams (g)	28.
pounds (lb)	kilograms (kg)	0.454
pints (pt)	liters (l)	0.47
quarts (qt)	liters (l)	0.95
gallons (gal)	liters (l)	3.79

ANSWERS TO SELECTED EXERCISES

CHAPTER ONE

EXERCISES 1.1

1. Quadrant I   3. none   5. Quadrant III   7. none

9. Quadrant IV   11. (2,3)   13. (−2,0)   15. (−5,2)   17. (5,0)

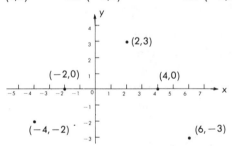

19. 3×2=6   Let {a,b,c} be the set of restaurants and {d,e} be the set of movies. Then we have
{(a,d),(a,e),(b,d),(b,e),(c,d),(c,e)}

21. {(3,x),(3,y),(3,z),(4,x),(4,y),(4,z)}   23. {(3,3),(3,4),(4,3),(4,4)}

25. •A×B={(1,1),(1,2),(2,1),(2,2),(3,1),(3,2)}   □B×A={(1,1),(1,2),(1,3),(2,1),(2,2),(2,3)}

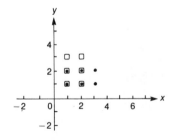

27. $A \times B = \{(x,y) \mid 0 < x \le 5 \text{ and } y = -1 \text{ or } 2\}$ $B \times A = \{(x,y) \mid x = -1 \text{ or } 2 \text{ and } 0 < y \le 5\}$

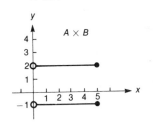 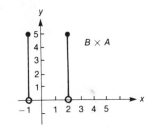

29. $A \times B = \{(x,y) \mid -5 < x < 1, \, y = 2\}$ $B \times A = \{(x,y) \mid x = 2 \text{ and } -5 < y < 1\}$

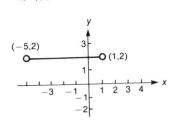 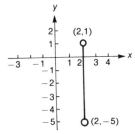

31. $A \times B = \{(x,y) \mid -2 < x < 2 \text{ and } y \le 4\}$ $B \times A = \{(x,y) \mid x \le 4 \text{ and } -2 < y < 2\}$

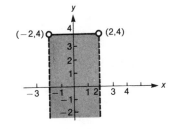 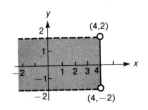

33. a. (34 N, 118.20 W)   b. (40 N, 75.10 W)   c. (30 N, 90.50 W)   d. (34 N, 81 W)

35. mk 37. k^2 39. k

EXERCISES 1.2

1. $\sqrt{41}/2$, $(2, \frac{3}{4})$ 3. $\sqrt{13}$, $(\frac{3}{2}, 6)$ 5. $\sqrt{2}$, $(\frac{1}{2}, -\frac{5}{2})$ 7. $\sqrt{137}$, $(\frac{5}{2}, -2)$

9. $\sqrt{2}$, $(1,0)$ 11. $\sqrt{74}/2$, $(-\frac{7}{4}, \frac{9}{4})$ 13. $[d(A,B)]^2 = [d(A,C)]^2 + [d(B,C)]^2$

15. $[d(B,C)]^2 = [d(A,C)]^2 + [d(A,B)]^2$ 17. $[d(B,C)]^2 = [d(A,B)]^2 + [d(A,C)]^2$

19. isosceles   21. isosceles   23. $(-28,13)$ 25. $(6,0)$, $\sqrt{18}$

27. $\sqrt{145}/2$, $\sqrt{130}/2$, $\sqrt{73}/2$

EXERCISES 1.3

1. function, $\{1,2,3,4\}$, $\{3,7,11,15\}$ 3. function, $\{4,6,8\}$, $\{2,4,6\}$

5. function, $\{0,3,4,5\}$, $\{2,8\}$ 7. $\{x \mid x \in R\}$ 9. $\{x \mid x \in R\}$

11. $\{t\,|\,t\in R,\ t\neq 0,1\}$ 13. $\{x\,|\,x\in R,\ x\neq\pm 1\}$

15. $\{x\,|\,-2\leq x\leq 2\}$ 17. $\{x\,|\,x\in R,\ x\neq -\frac{1}{2}\}$ 19. $\{x\,|\,x\in R,\ x\neq 1\}$

21. a. -1 b. 0 c. 0 d. $-\frac{3}{4}$ 27. a. ± 1 b. ± 2 c. $\pm\sqrt{5}$

23.

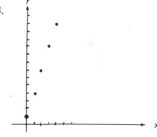

25.

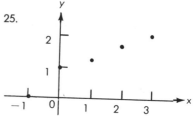

29. a. even   b. odd   c. even   d. neither   31. $f(x)=x+x^2$

33.

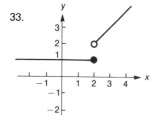

35.

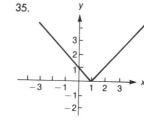

37. b.

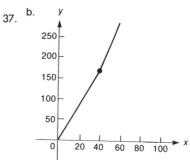

39. $1,4,9,16,25$ 41. a. $C(x)=8+0.12x$ b. $17

EXERCISES 1.4

1. $(f+g)(x)=2x^2+3x+1,$ domain $f+g=\{x\,|\,x\in R\}$
    $(f-g)(x)=3x+1-2x^2,$ domain $f+g=\{x\,|\,x\in R\}$
     $(fg)(x)=6x^3+2x^2,$ domain $fg=\{x\,|\,x\in R\}$
     $(f/g)(x)=(3x+1)/(2x^2),$ domain $f/g=\{x\,|\,x\in R,\ x\neq 0\}$
     $(g/f)(x)=(2x^2)/(3x+1),$ domain $g/f=\{x\,|\,x\in R,\ x\neq -\frac{1}{3}\}$

3. $(f+g)(x)=x^2+2x-1+\sqrt{x^2-1}$
    $(f-g)(x)=x^2+2x-1-\sqrt{x^2-1}$
     $(fg)(x)=(x^2+2x-1)\,\sqrt{x^2-1},$ domain of $f+g,\ f-g,\ fg$ is $\{x\,|\,x\geq 1\}\cup\{x\,|\,x\leq -1\}$
     $(f/g)(x)=(x^2+2x-1)/\sqrt{x^2-1}$ domain $f/g=\{x\,|\,x>1\}\cup\{x\,|\,x<-1\}$
     $(g/f)(x)=(\sqrt{x^2-1})/(x^2+2x-1),$ domain $g/f=\{x\,|\,x\geq 1\}\cup\{x\,|\,x\leq -1,\ x\neq -1-\sqrt{2}\}$

5. $(f+g)(x)=(1/x^2)+(1/\sqrt{x})$
    $(f-g)(x)=(1/x^2)-(1/\sqrt{x})$
     $(fg)(x)=1/x^{5/2}$ domain of $f+g,\ f-g,\ fg$ is $\{x\,|\,x>0\}$
     $(f/g)(x)=\sqrt{x}/x^2=1/x^{3/2},$ domain $f/g=\{x\,|\,x>0\}$
     $(g/f)(x)=x^2/\sqrt{x}=x^{3/2},$ domain $g/f=\{x\,|\,x>0\}$

7. $(f\cdot g)(x)=2x+1,\ (g\cdot f)(x)=2x+2,$ domain $f\cdot g$ and $g\cdot f$ is $\{x\,|\,x\in R\}$

9. $(f\cdot g)(x)=x+2,$ domain $f\cdot g=\{x\,|\,x\geq -1\};$ $(g\cdot f)(x)=\sqrt{x^2+2},$ domain $g\cdot f=\{x\,|\,x\in R\}$

11. $(f\cdot g)(x)=[x/(x+1)],$ domain $f\cdot g=\{x\,|\,x\in R,\ x\neq -1\};$
     $(g\cdot f)(x)=1/x,$ domain $g\cdot f=\{x\,|\,x\in R,\ x\neq 0\}$

13. $(f\cdot g)(x)=1/(x^2+3),$ domain $f\cdot g=\{x\,|\,x\in R\};$
     $(g\cdot f)(x)=(\sqrt{1+3x^4})/x^2,$ domain $g\cdot f=\{x\,|\,x\in R,\ x\neq 0\}$

15. $1+\sqrt{5}$ 17. $\sqrt{5}$ 19. $2\sqrt{5}$

21. $(f \circ f)(x) = 4x + 3$,   domain $f \circ f = \{x \mid x \in R\}$

23. $(f \circ f)(x) = (2x+1)/(1+x)$, domain $f \circ f = \{x \mid x \in R,\ x \neq 0,\ x \neq -1\}$

EXERCISES 1.5

1. $f^{-1} = \{(3, 0.5)(7, -5)(4, 17)\}$ 3. f is not one-to-one

5.   Select $x_1 = 1$, $x_2 = -1$. We find $(f)(x_1) = f(x_2)$ even though $x_1 \neq x_2$.

7. $f^{-1}(x) = \tfrac{1}{7}(x-2)$ 9. $f^{-1}(x) = x^2 - 2,\ x \geq 0$ 11. $f^{-1}(x) = \sqrt{x/3},\ x \geq 0$

13. $f^{-1}(x) = \sqrt[3]{x-1}$ 15. $f^{-1}(x) = (x-5)^3$ 17. a.   5   b. -135

REVIEW OF CHAPTER ONE

2.   yes   3.   no   5. $-1, 3,$ II   7.   no, $A = \{1\}$, $B = \{1\}$, $A \times B = B \times A = \{(1,1)\}$

9. $\sqrt{58}$ 11. $\tfrac{5}{2}, 1$ 13. $\{x \mid x \geq 2,\ x \neq 3\}$ 15. $-\tfrac{7}{2}$

17. $2(4x^3 - x^2 - 2x + 1)/(1 - 4x)$ 19.   1   29. $(x+7)/6$ 31.   u, v

TEST ONE

1. $-3, 5, 2$ 2. $A \times B = \{(x,y) \mid -2 \leq x \leq 3,\ y = 2\}$ 3.   a. $2\sqrt{13}$ b. $-1, 1$

4.   a. $\{-5, -4, -3, -2\}$,   b. $\{1, 2, 3, 4\}$ c.   Yes. No two ordered pairs have the same first component.

5.   a.   3   b. $\tfrac{7}{2}$ c.   4   6. $\{x \mid x \in R,\ x \neq 1, 4\}$ 7. $\{x \mid x \leq 2\}$

8. ± 2 9.   a. $(x\sqrt{x-1}+1)/x$ b. $(x\sqrt{x-1}-1)/x$ c. $\sqrt{x-1}/x$ d. $\sqrt{2}/6$ e.   2

10. a. $x + 1,\ x \geq 0$ b. $\sqrt{x^2 + 1}$ c. 2   d.   2   11.   Yes. $f(x_1) = f(x_2) \rightarrow x_1 = x_2$

12. $g^{-1}(x) = x^2 + 1,\ x \geq 0$

CHAPTER TWO

EXERCISES 2.1

1.

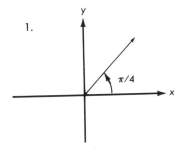

3.

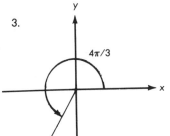

5.

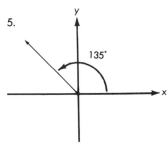

7.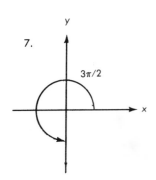

9.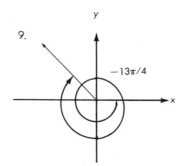

11.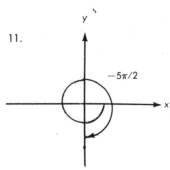

13. 0.49 rad   15. 12.57 rad   17. 0.85 rad   19. 45°   21. −405°

23. −900°   25. 420°, −300°, −660°   27. 210°, 570°, −510°

29. 600°, −120°, −480°   31. −30°, 330°, 690°   33. 189°, 549°, −531°

35. 50.26 cm   37. 2.25 rad   39. 5419 km   41. 3350 km 43. 455

EXERCISES 2.2

1. 32 cm/sec   3. 6 mm/sec   5. 9.8 m/min   7. 2 rad/sec
9. 16.77 rad/sec   11. 50 rad/sec   13. 50240 cm/min
15. a. 70.4 rad/sec b. 11.2 rev.

17. 200 ft/sec   19. 200 m/min   21. 1675 km/hr

EXERCISES 2.3

1. II   3. IV   5. $-\sqrt{2}/2, \sqrt{2}/2$ 7. −1,0   9. $\sqrt{3}/2$, ½

11. $(-\sqrt{2})/2, \sqrt{2}/2$ 13. $\sqrt{3}/2$, ½   15. $-\sqrt{3}/2, -$ ½ 17. −5/13, 12/13

19. −4/5, 3/5   21. 8/17, −15/17 23. −15/17, −8/17

25. 4/5   27. −12/13   29. $-\sqrt{3}/2$ 31. 1

33. $\sqrt{1-m^2}$ 35. $\sqrt{1-m^2}$ 37. 2π 39. π

EXERCISES 2.4

1. $y = 2 \cos \theta$, amplitude 2, period 2π; no phase shift.

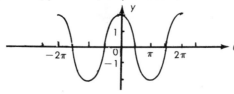

3. $y = -2 \cos \theta$, amplitude 2, period 2π; no phase shift.

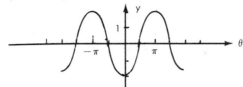

5. $y = \frac{1}{2} \sin 2\theta$, amplitude $\frac{1}{2}$, period π; no phase shift.

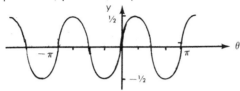

7. $y = -3 \sin \frac{1}{2}\theta$, amplitude 3, period 4π; no phase shift.

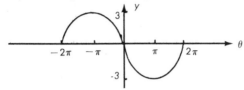

9. $y = \cos \frac{1}{2}\theta$, amplitude 1, period 4π; no phase shift

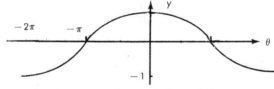

11. $y = \frac{1}{4} \sin \theta$, amplitude $\frac{1}{4}$, period 2π; no phase shift.

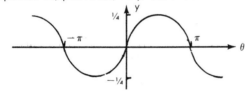

13. $y = 2 \cos(\theta/4)$, amplitude 2, period 8π; no phase shift.

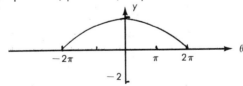

15. $y = \sin(-3\theta)$, amplitude 1, period $2\pi/3$; no phase shift.

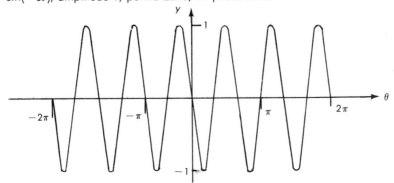

17. $y = -\frac{1}{2}\cos 3\theta$, amplitude $\frac{1}{2}$, period $(2\pi/3)$; no phase shift.

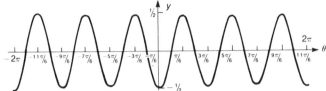

19. $y = \sin(\theta + \pi)$, amplitude 1, period 2π; phase shift $-\pi$.

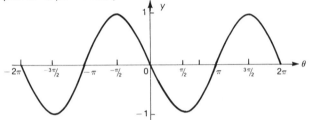

21. $y = 2\cos[\theta - (\pi/2)]$, amplitude 2, period 2π; phase shift $\pi/2$.

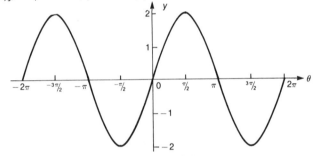

25. $y = 3\sin[\theta + (\pi/6)]$, amplitude 3, period 2π; phase shift $-\pi/6$.

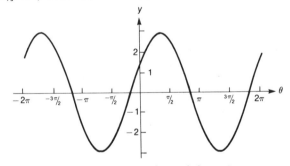

27. $y = 2\sin[2\theta - (\pi/6)]$, amplitude 2, period π; phase shift $\pi/12$.

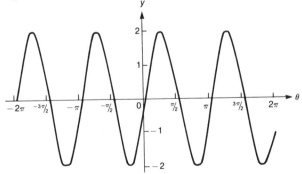

29. Amplitude 3, period 4π and phase shift π.

31. $y = \sin\theta + \cos\theta$

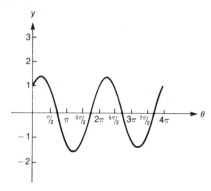

EXERCISES 2.5

1. $\cos\theta \neq 0$

3. $\sin\theta \neq 0$

5. $\sin\theta = -8/17$
   $\cos\theta = -15/17$
   $\cot\theta = 15/8$
   $\sec\theta = -17/15$
   $\csc\theta = -17/8$

7. $\sin\theta = 12/13$
   $\tan\theta = 12/5$
   $\cot\theta = 5/12$
   $\csc\theta = 13/12$
   $\sec\theta = 12/5$

9. $\cos\theta = 5/\sqrt{4}$
   $\sin\theta = -3/\sqrt{34}$
   $\tan\theta = -3/5$
   $\csc\theta = -\sqrt{34}/3$
   $\sec\theta = \sqrt{34}/5$

11. $\cos\theta = -3/5$
    $\sin\theta = -4/5$
    $\tan\theta = 4/3$
    $\csc\theta = -5/4$
    $\sec\theta = -5/3$

13. $\cos\theta = -\sqrt{5}/3$
    $\tan\theta = -2/\sqrt{5}$
    $\cot\theta = -\sqrt{5}/2$
    $\csc\theta = 3/2$
    $\sec\theta = -3/\sqrt{5}$

15. $\sin\theta = -3/\sqrt{10}$
    $\cos\theta = -1/\sqrt{10}$
    $\cot\theta = 1/3$
    $\csc\theta = \sqrt{10}/3$
    $\sec\theta = \sqrt{10}$

17. $\cos\theta = 1/5$
    $\sin\theta = -(2\sqrt{6})/5$
    $\tan\theta = -2/\sqrt{6}$
    $\cot\theta = -\sqrt{6}/12$
    $\csc\theta = -(5\sqrt{6})/12$

19. $\cos\theta = 8/17$
    $\sin\theta = -15/17$
    $\tan\theta = -15/8$
    $\cot\theta = -8/15$
    $\sec\theta = 17/8$

21. $\dfrac{\sin\theta}{\cos\theta}$

23. $\dfrac{\sin\theta}{\cos^2\theta}$

25. $(\sin^2\theta - \cos^2\theta)/(\cos^2\theta\ \sin^2\theta)$

27. $1/\cos^2\theta$

29. $(\cos^4\theta - \sin^4\theta)/(\sin^2\theta\ \cos^2\theta)$

31.

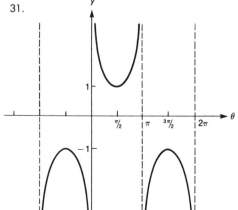

35.

43.

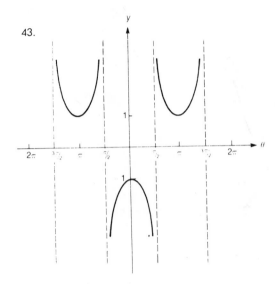

47. 0.2022, 0.9793, 0.2065

49. 0.9367, -0.3502, -2.675

51. -0.8107, -0.5854, 1.385

53. 0.6947, 0.7193, 0.9657

55. 0.3137, 0.9495, 0.3304

57. -0.9350, 0.3546, -2.6371

EXERCISES 2.6

1. $(\frac{1}{2}, \sqrt{3}/2)$ 3. $(\sqrt{3}/2, \frac{1}{2})$ 5. $(-\sqrt{2}/2, -\sqrt{2}/2)$ 7. $(0, -1)$

9. $(-\sqrt{3}/2, \frac{1}{2})$ 11. $(-\frac{1}{2}, -\sqrt{3}/2)$ 13. $\sin 0.34 = 0.3335$, $\cos 0.34 = 0.9427$,
                                                                $\tan 0.34 = 0.3537$, $\csc 0.34 = 2.998$,
                                                                $\sec 0.34 = 1.061$, $\cot 0.34 = 2.827$

15. -0.8674, -1.534 19. 1,   1   23. $-\frac{1}{2}$, -2
    0.4975,   1.319   0,   undef. $\sqrt{3}/2$, $2/\sqrt{3}$
    -1.743, -1.163 undef.,   0 $-\sqrt{3}/3$, $-\sqrt{3}$

25. 1,   1   27. $-\sqrt{2}/2$, -2 29. $-\frac{1}{2}$, -2 31. $3/\sqrt{10}$, $\sqrt{10}/3$
    0,   undef. $\sqrt{2}/2$, $\sqrt{2}$ $-\sqrt{3}/2$, $-2\sqrt{3}/3$ $-1/\sqrt{10}$, $-\sqrt{10}$
    undef., 0 -1, -1 $\sqrt{3}/3$, $\sqrt{3}$ -3, $-1/3$

33. $-1/\sqrt{5}$, $-\sqrt{5}$ 35. $-2/\sqrt{13}$, $-\sqrt{13}/2$
    $2/\sqrt{5}$, $\sqrt{5}/2$ $3/\sqrt{13}$, $\sqrt{13}/3$
    $-1/2$, -2 $-2/3$, $-3/2$

39. $-5/3$, 5/4, $-4/3$ 41. $-\sqrt{5}/2$, $-\sqrt{5}$, 1/2

43. -6.7 cm, 3.19 cm   45. 0.34

47. a. -0.029 b. 0 c. 0 d. 0

49. 5, $(2\pi)/377$, $377/(2\pi)$, $(60\pi)/377$ 51. 0.245π

REVIEW OF CHAPTER TWO

6. 0.4638   7. 77.92°   9. −320°   12. 2262 ft/min

15. a. $\sqrt{3}/2, -\frac{1}{2}, \sqrt{3}$ b. $-\sqrt{2}/2, \sqrt{2}/2, -1$ c. $-1, 0$, undef.

16. $\sin\theta = -4/5$, $\cos\theta = 3/5$, $\tan\theta = -4/3$

TEST TWO

1.   a. 0.56, I   b. −2.36, III   c. 2.91, II

2.   a. 450°,   b. 540° 3.   2.25   4.   37.88 rad/sec

5. $\sin\theta = 3/\sqrt{13}$, $\cos\theta = -2\sqrt{13}$, $\tan\theta = -3/2$, $\csc\theta = \sqrt{13}/3$, $\sec\theta = -\sqrt{13}/2$, $\cot\theta = -2/3$

6.   a. $\sqrt{2}/2$ b. -1 c. $-\frac{1}{2}$

7. $\sin\theta = -3/5$, $\tan\theta = -3/4$, $\csc\theta = -5/3$, $\sec\theta = 5/4$, $\cot\theta = -4/3$

9. $\sin t = -\sqrt{3}/2$, $\cos t = \frac{1}{2}$, $\tan t = -\sqrt{3}$, $\csc t = -2/\sqrt{3}$, $\sec t = 2$, $\cot t = -1/\sqrt{3}$

11. 0.3090

CHAPTER THREE

EXERCISES 3.1

1. $(\sqrt{2}/4)(1+\sqrt{3})$ 3. $(\sqrt{2}/4)(\sqrt{3}-1)$ 5. $(3+\sqrt{3})/(3-\sqrt{3})$

7. $(-\sqrt{2}/4)(\sqrt{3}-1)$ 9. $(\sqrt{3}-3)/(\sqrt{3}+3)$ 11. $(\sqrt{2}/4)(1+\sqrt{3})$

13. $(-\sqrt{2}/4)(\sqrt{3}+1)$ 15. $(\sqrt{2}/4)(1+\sqrt{3})$ 17. $(\sqrt{2}/4)(\sqrt{3}-1)$

21. 63/65   23. 16/65   25. 52/33

27. 13/85   29. 84/85   31. −77/36

47. $2\cos 2\theta \cos\theta$ 49. $2\sin(\alpha-\beta)^{-1}\cos(\alpha+\beta)^{-1}$

EXERCISES 3.2

1. $(\sqrt{2+\sqrt{2}})/2$ 3. $(\sqrt{2}/4)(\sqrt{3}-1)$ 5. $(-\sqrt{2}/4)(\sqrt{3}-1)$

7. $2/(\sqrt{2-\sqrt{2}})$ 9. $(3-\sqrt{3})/(3+\sqrt{3})$ 11. $\left(\dfrac{4-\sqrt{2}(1+\sqrt{3})}{4+\sqrt{2}(1+\sqrt{3})}\right)^{1/2}$

13. $\left(\dfrac{4-\sqrt{2}(1+\sqrt{3})}{4+\sqrt{2}(1+\sqrt{3})}\right)^{1/2}$ 19. $2\cos\theta$ 21. $\sin 6\theta$

23. $\cos 2\theta$ 25. $\tan\theta$ 27. $1/3$, $(-2\sqrt{2})/3$

29. $\frac{1}{4}(8-3\sqrt{7})^{1/2}$, $-\frac{1}{4}(8+3\sqrt{7})^{1/2}$ 31.   −24/25, −7/25, 24/7

EXERCISES 3.3

1. $\pi/2$

3. $\pi/6,\ 11\pi/6$

5. $\pi/4,\ 5\pi/4$

7. $5\pi/4,\ 7\pi/4$

9. $2\pi/3,\ 4\pi/3$

11. $5\pi/6,\ 11\pi/6$

13. $s=\{\theta\,|\,\theta=(\pi/6)+2n\pi\ \text{or}\ \theta=(5\pi/6)+2n\pi,\ n\in Z\}$

15. $s=\{\theta\,|\,\theta=n\pi,\ n\in Z\}$

17. $\pi/3$

19. 0

23. π

25. $-\pi/4$

27. $3\pi/4$

29. $\pi/2$

31. $-\pi/2$

33. $3\pi/4$

35. $-\pi/4$

37. $\sqrt{2}/2$

39. $4/5$

41. $3/\sqrt{11}$

43. $\sqrt{3}$

45. $-\sqrt{2}/2$

47. $24/25$

49. $-[(2\sqrt{5})/25]$

51. 1

53. $(7\sqrt{2})/10$

55. $\sqrt{3}/2$

57. $-7/25$

59. Let $\text{Cos}^{-1}x=u$, then $\text{Cos}\ u=x$, $\sin u=\sqrt{1-x^2}$ or $\sin(\text{Cos}^{-1}x)=\sqrt{1-x^2}$.

EXERCISES 3.4

1. $\{\theta\,|\,\theta=(2\pi/3)+2n\pi,\ \text{or}\ \theta=(4\pi/3)+2n\pi,\ n\in Z\}$

3. $\{\theta\,|\,\theta=(\pi/3)+2n\pi\ \text{or}\ \theta=(2\pi/3)+2n\pi,\ n\in Z\}$

5. $\{0,\ \pi,7\pi/6,\ 11\pi/6\}$

7. $\{\pi/6,\ 5\pi/6,\ 7\pi/6,\ 11\pi/6\}$

9. $\{0,\ \pi/4,\ \pi,\ 5\pi/4\}$

11. $\{0,\ 2\pi/3,\ \pi,\ 4\pi/3\}$

13. $\{0,\ \pi/6,\ 5\pi/6,\ \pi\}$

15. $\{\pi/4,\ 2\pi/3,\ 5\pi/4,\ 4\pi/3\}$

17. $\{\pi/6,\ 5\pi/6,\ \pi\}$

19. $\{\pi/12,\ 13\pi/12\}$

21. $\{\pi/2,\ 3\pi/2\}$

23. $\{\pi/6,\ 5\pi/6,\ 7\pi/6,\ 11\pi/6\}$

25. $\{0,\ 5\pi/6,\ \pi,\ 7\pi/6\}$

27. $\{5\pi/6,\ 11\pi/6\}$

29. $0.2022,\ 0.9793,\ 0.2065$

31. $0.4210,\ -0.9070,\ -0.4642$

33. $0.4576,\ 0.8892,\ 0.5147$

35. $0.5331,\ 0.8460,\ 0.6301$

37. $17°0'$

39. $41°13'$

41. $\{63°26',\ 243°26',\ 14°28',\ 165°32'\}$

43. $\{205°26',\ 334°34'\}$

45. $\{66°25',\ 131°49',\ 228°11',\ 293°35'\}$

REVIEW OF CHAPTER THREE

4. a. $\{\theta\,|\,\theta=(\pi/3)+2n\pi\ \text{or}\ \theta=(2\pi/3)+2n\pi,\ n\in Z\}$ b. $2\pi/3$
   c. $\{\theta\,|\,\theta=(3\pi/4)+2n\pi\ \text{or}\ \theta=(7\pi/4)+2n\pi,\ n\in Z\}$ d. $-\pi/4$

8. a. $3/5$ b. $\pi/6$ c. $\sqrt{22}/11$

9. a. $\{0°, 180°, 210°, 330°\}$ b. $\{30°, 150°, 210°, 330°\}$ c. $\{14°29', 165°31', 210°, 330°\}$

   d. $\{0°, 120°, 240°\}$ e. $\{135°, 315°\}$ f. $\{60°, 131°49', 228°11', 330°\}$

   g. $\{46°37', 106°51', 253°9', 313°23'\}$

TEST THREE

1. a. $(\tan\alpha+\tan\beta)/(1+\tan\alpha\tan\beta)$ b. $(\sqrt{2}/4)(1+\sqrt{3})$ c. 0.4067

2. $-7/25$ 3. $(\sqrt{2-\sqrt{2}})/2$ 4. $2\tan\theta$

6. $\{\theta\,|\,\theta=(\pi/6)+2n\pi \text{ or } \theta=(7\pi/6)+2n\pi,\ n\in Z\}$ 7. $20°12'$

8. $\pi/3$ 9. $5/6$ 10. 1

11. a. $\{\pi/6, \pi/2, 5\pi/6, 3\pi/2\}$ b. $\{\pi/6, 5\pi/6\}$ c. π d. $\{205°26', 334°34'\}$

CHAPTER FOUR

EXERCISES 4.1

1. $a=22.79, b=66.19, B=71°$ 3. $c=10, A=36.87°, B=53.13°$

5. $a=0.3948, c=9.71, A=2°20'$ 7. $b=668.58, c=681.88, A=11°20'$

9. $a=0.0307, b=0.0325, A=43°20'$ 11. $b=31.31, B=64.4°, c=34.72$

13. $a=399.07, c=474.23, A=57.3°$ 15. $c=15, b=9, A=53.13°, B=36.87°$

17. 127.2 meters 19. $36.87°$ 21. 3.57 meters 23. 29.02 meters

25. 513 meters 27. 25.4 meters 29. 21.33 meters, 35.22 meters

31. 5.06 kilometers 33. 29.87 kilometers, 33.42 kilometers

EXERCISES 4.2

1. $a=6.39, c=8.21, C=40°$ 3. $C=110°, b=13.44, c=16.48$

5. $b=5.05, c=4.29, C=33.3°$ 7. $c=3.46, B=90°, C=60°$

9. no solution

11. two solutions: $c=28.5, A=41.81°, C=108.19°$ or $c=6.15, A=138.19°, C=11.81°$

13. $A=70°32', C=64°28', c=3.83$ 15. $B=40°14', A=88°6', a=13$

19. 73.67 meters 21. 10.11 km 23. 116.19 meters 25. 50 km

27. $12.5°$

EXERCISES 4.3

1. $A=30°$, $B=61°1'$, $C=88°59'$

3. $A=46°34'$, $B=104°29'$, $C=28°57'$

5. $A=106°54'$, $B=32°42'$, $C=40°24'$

7. $b=14.87$, $A=23°48'$, $C=126°12'$

9. $a=26.69$, $B=41°39'$, $C=28°20'$

11. $b=13$, $A=22°37'$, $C=22°23'$

13. 5.33   15. 31.30   17. 35.12   19. 61.95

21. $58°27'$, $121°33'$, no   23. 291.63 km   25. $55°55'$

27. $36.20°$, $117.16°$, $62.44°$, $143.40°$; 49,77

REVIEW OF CHAPTER FOUR

1. a,b,c,A,B,C

3. $\sin A=a/c$, $\cos A=b/c$, $\tan A=a/b$, $\csc A=c/a$, $\sec A=c/b$, $\cot A=b/a$

4. $180°$ 5. angle of elevation   6. angle of depression

7. $A=37°50'$, $B=52°10'$, $b=5.92$ 8. 603.48   9. oblique triangle

12. $B=45.58°$, $C=104.42°$, $c=96.85$ 13. 36.4   15. ½ bc sin A

17. $b=11.67$, $A=28°26'$, $C=39°24'$ 19. $\sqrt{s(s-a)(s-b)(s-c)}$ 20. $100\sqrt{3}$

TEST FOUR

1. 30.9   2. $B=22.02°$, $C=127.58°$, $c=31.7$

3. $A=29°41'$, $B=82°4'$, $C=68°15'$ 4. 19.81   5. 173.2, 200

CHAPTER FIVE

EXERCISES 5.1

1. $3+7i$ 3. $1+4i$ 5. $6+4i$ 7. $1-7i$ 9. $-2+2i$

11. $-3/4i$ 13. $-7+6i$ 15. 5   17. $-2+10i$ 19. 29

21. $-5+12i$ 23. $-2-2i$ 25. $5/2+½i$ 27. $-i$ 29. $1-i$

31. $⅓+(2\sqrt{2}/3)i$ 33. $6/13-(9/13)i$ 35. $-1+i$ 37. $-4i$ 39. $-½(1+i)$

41. $1-2i$ 43. $½(3-7i)$ 45. $1/10(29-7i)$ 51. $x=-3$, $y=-5/2$

57. a. 1   b. i c. -1

EXERCISES 5.2

1. $1-i$ 3. 4   5. $10-11i$ 7. $10-11i$ 13. $\sqrt{17}$

15. $\sqrt{2}/2$ 17. 5   19. $1/5(3+4i)$ 21. $1/13(18+i)$ 23. $1/26(5+6i)$

25. -2 27. $5-2i$ 29. $1/5(19-18i)$

EXERCISES 5.3

15. $(2,\ \pi/3)$ 17. $(2,\ \pi/2)$ 19. $(4,\ \pi)$ 21. $(\sqrt{6},\ 5\pi/4)$ 23. $r^2=\csc\theta\sec\theta$

25. $r=2\cos\theta$ 27. $r=4\cot\theta\csc\theta$ 29. $(x^2+y^2-x)^2=x^2+y^2$

31. $x=4$

35.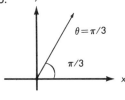

37.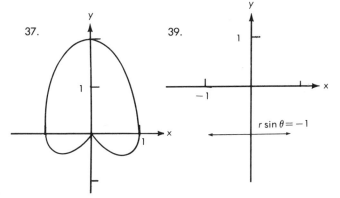

39.

41.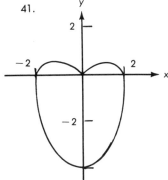

43.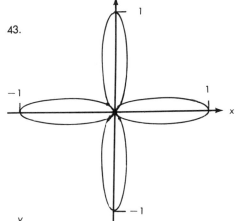

45.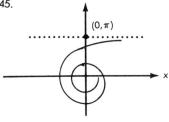

47.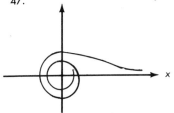

49. $(2,0),\ (2,\pi)$

EXERCISES 5.4

1. $\sqrt{2}$ cis $3\pi/4$ 3. 2 cis $\pi/3$ 5. 4 cis $11\pi/6$ 7. 2 cis $\pi/2$

9. 17 cis 0   11. $3\sqrt{2}$ cis $5\pi/4$ 13. 2, i 15. 20, $1/10(1-\sqrt{3}i)$

17. $4+4\sqrt{3}i$, $-\frac{1}{4}(1-\sqrt{3}i)$ 19. 1, -1 21. $16+16i$

23. $8i$ 25. -64 27. $-81/2(1+\sqrt{3}i)$ 29. $-\frac{1}{2}(1-i)$

31. $-\frac{1}{2}(\sqrt{6}-\sqrt{2}i)$, $\frac{1}{2}(\sqrt{6}-\sqrt{2}i)$ 33. $\sqrt{3}+i$, $-\sqrt{3}-i$

35. Let $a=\sqrt{2+\sqrt{3}}$ and $b=\sqrt{2-\sqrt{3}}$. Then the roots are: $a+bi$, $-b+qi$, $-a-bi$, $b-ai$

. 37. $(\sqrt{2}/2)(1+i)$, $(\sqrt{2}/2)(-1+i)$, $-(\sqrt{2}/2)(1+i)$, $(\sqrt{2}/2)(1-i)$

REVIEW OF CHAPTER FIVE

3. If $z=x+iy$, then $\bar{z}=x-iy$ and $|z|=\sqrt{x^2+y^2}$

4. $3+2i$ 5. $-2+i$ 6. $14-2i$ 7. $1/10(11-3i)$

8. i 9. $1/13(3+2i)$ 10. $5+5i$ 11. $x=3$, $y=-5$

12. yes, yes, no   13. 5   16. a. $\pi/6$ b. 2

18. a. $(-3/2)+(3\sqrt{3}/2)i$ b. $(-5/2)+(5\sqrt{3}/2)i$ c. $(3/2)-(3\sqrt{3}/2)i$
     d. $(5/2)+5\sqrt{3}/2)i$

19. $(2, \pi/6)$, $(2, 7\pi/6)$ 20. $r=-6 \sin \theta$ 21. $\sqrt{x^2+y^2}-x-1=0$ 23. cis $11\pi/6$

24. $1+\sqrt{3}i$ 25. $-6\sqrt{2-\sqrt{3}}+i\,6\sqrt{2+\sqrt{3}}$

26. 2 cis $\pi/12$ 27. $z^n=r^n$ cis $n\,\theta$

28. 64 cis π, cis 50π, 2^{12}cis 12π 29. $x^{1/n}=r^{1/n}$cis θ/n

30. $\sqrt{3}/2+\frac{1}{2}i, \sqrt{3}/2-\frac{1}{2}i, -i$

TEST FIVE

1. a. $7-4i$ b. $24-10i$ c. $2\sqrt{13}$ d. $24+10i$ 2. $1/53(41+11i)$

3. a. $(1, -\sqrt{3})$ b. $(6, 5\pi/3)$

4. a. 2 cis $(-\pi/6)$, 2 cis $\pi/3$ b. 4 cis $\pi/6$, cis $(-\pi/2)$

5. a. $-64\sqrt{3}-64i$ b. $81/2(1+\sqrt{3}i)$ 6. $2\sqrt{2}$ cis$(-\pi/12)$, $2\sqrt{2}$ cis $7\pi/12$, $2\sqrt{2}$ cis $15\pi/12$

CHAPTER SIX

EXERCISES 6.1

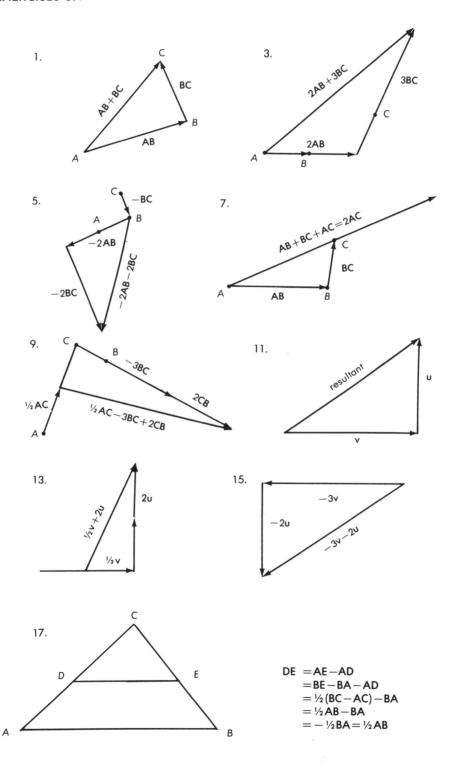

1.

3.

5.

7.

9.

11.

13.

15.

17.

DE = AE − AD
 = BE − BA − AD
 = ½(BC − AC) − BA
 = ½AB − BA
 = −½BA = ½AB

EXERCISES 6.2

1. $3\sqrt{2}$, $\pi/4$ 3. $4\sqrt{2}$, $3\pi/4$ 5. 5, $\pi/2$ 7. $<3,-1>$, $<1,7>$

9. $<6,1>$, $<-6,-9>$ 11. $<-5,-5>$, $<1,3>$

13. $<6,6>$, $<4,-4>$ 15. $<0,8>$ 17. $<10,4>$

19. $<-24,3>$ 21. $<12,1>$ 23. $<23,-11>$ 25. $\sqrt{13}+\sqrt{34}$

27. $\sqrt{205}$ 29. $\sqrt{865}$ 35. $<5,36°52'>$

37. $<5\sqrt{13},146°46'>$ 39. $2i+2\sqrt{3}j$

EXERCISES 6.3

1. a. $21\sqrt{3}$ b. 21   c. $-21\sqrt{2}$ d. $-21\sqrt{3}$

7. 1   9. -2 11. 4   13. -9 15. 90°

17. 143°8'   19. 53°8'   29. 327.26, 20°33'

REVIEW OF CHAPTER SIX

2. $\sqrt{34}$ 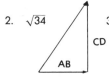   3. x component, y component   4. $\sqrt{c^2+d^2}$

5. $3\sqrt{2}$, $7\pi/4$ 6. a. $<ka,kb>$ b. $<a+c,b+d>$ c. 0   d. u

7. a. $<2,-3>$ b. $<2,-4>$ c. $<-4,7>$ d. $<-11,19>$ e. $\sqrt{5}$ f. $\sqrt{13}$

8. $u=ai+bj$

9. a. $\sqrt{10}$ b. $5i+4j$ c. $\sqrt{10}+\sqrt{29}$ d. $\sqrt{41}$ e. $i-6j$ f. $-4i-27j$

10. $<2\sqrt{2},3\pi/4>$ 11. $<-\sqrt{3},1>$ 13. yes   14. perpendicular

15. $-15\sqrt{2}$ 16. 1   17. 3/2   19. $\pi/2$

TEST SIX

1. a.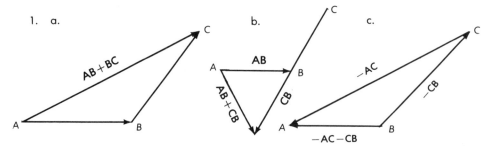

2. $-5/2, 5\sqrt{3}/2$

3. a. 5   b. $<0,1>$ c. $<-8,5>$ d. $<-20,13>$ e. 1   f. $5-2\sqrt{5}$ g. 25

4. a. 2   b. 12   c. 34   d. 14   5. $85°36'$

6. 15/8   7. $1/5(4i+3j)$ 8. $10, 10\sqrt{3}$ 9. 36.06, S 36°7'W

CHAPTER SEVEN

EXERCISES 7. 2

1. $x^2+y^2+2x-4y-11=0$ 3. $x^2+y^2+4x+8y-5=0$ 5. $x^2+y^2+x-y-(9/2)=0$

7. $(3,0), r=5$ 9. $(-2,3), r=6$ 11. $x^2+y^2-6x+8y-72=0$

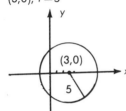
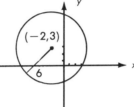
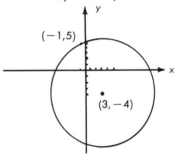

13. $x^2+y^2+2x-6y-6=0$ 15. $x^2+y^2-2x-4y-3=0$

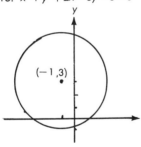
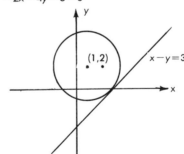

17. $144x^2+144y^2-1104x-168y-15175=0$ $C(23/6, 17/12), r=\sqrt{1445}/12$

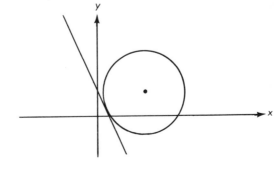

19. $x^2+y^2-10x-10y+25=0$ 21. $x^2+y^2-4x-5=0$ 23. $(7/6, 7/6), r=\sqrt{50}/6$

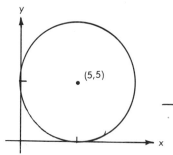

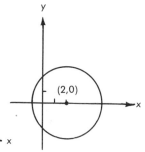

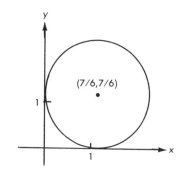

25. $(-2,1), r=5$

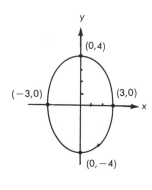

EXERCISES 7.3

1. major axis 8
    minor axis 6
    foci $(0, \pm\sqrt{7})$
    vertices $(0, \pm4)$
    $e=\sqrt{7}/4$

3. major axis 10
    minor axis 8
    foci $(\pm3,0)$
    vertices $(\pm5,0)$
    $e=3/5$

5. major axis $2\sqrt{10}$
    minor axis 4
    foci $(0, \pm\sqrt{6}$
    vertices $(0, \pm10)$
    $e=(6/10)^{1/2}$

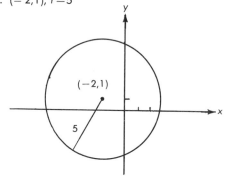

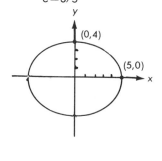

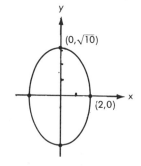

7. major axis 6
    minor axis $2\sqrt{3}$
    foci $(0, \pm 6)$
    vertices $(0, \pm 3)$
    $e = \sqrt{2}/3$

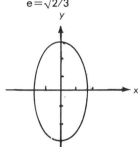

9. major axis 8
    minor axis 4
    foci $(\pm\sqrt{12}, -2)$
    vertices $(5, -2)$, $(-3, -2)$
    $e = \sqrt{3}/2$

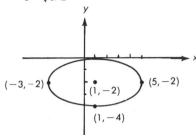

11. not an ellipse

13. $16x^2 + 9y^2 = 144$

15. $25x^2 + 9y^2 = 225$

17. $9x^2 + 16y^2 = 225$

19. $x^2 + 2(y-1)^2 = 2$

21. $9x^2 + 25y^2 = 225$

23. $x^2 + 3y^2 = 24$

25. $16x^2 + 12y^2 = 147$

27. a. 91.32 million miles   b. 94.48 million miles

29. 0.8182

EXERCISES 7.4

1. $y^2 = -8x$

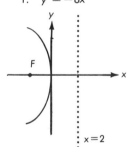

3. $x^2 = -16y$

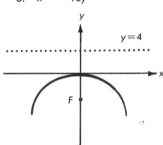

5. $y^2 = -10x$

7. $y^2 = -12x$

9. $y^2 = 8x$

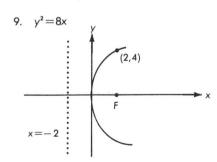

11. $V = (0, -2)$ principal axis; y-axis. $x^2 = 12(y-2)$

13. $V = (2,2)$, principal axis; $y = 2$. $(y-2)^2 = 12(x-2)$

15. $V = (2,4)$, principal axis; $x = 2$. $(x-2)^2 = -4(y-4)$

17. $V=(-3/2, 1)$, principal axis; $y=1$. $(y-1)^2=10(x+3/2)$

19. $(x-1)^2=-8(y-7/8)$, $V=(1,7/8)$, $F=(1,-9/8)$, directrix $y=23/8$

21. $y^2=-4(x-2)$, $V=(2,0)$, $F=(1,0)$, directrix $x=3$

23. $(x+5)^2=-\frac{1}{4}(y+6)$,•$V=(-5,-6)$, $F=(-5,-97/16)$, directrix $y=-95/16$

25. $(y-2)^2=-2(x+3/2)$, $V=(-3/2, 2)$, $F=(-2,2)$, directrix $x=-1$

27. $u^2=6v$, $x=u$, $y-3/2=v$.   29. $u^2=12v$, $x+3=u$, $y-4=v$

EXERCISES 7.5

1. Center $(0,0)$, Foci $(5,0)$; $(-5,0)$
    $e=5/4$, Vertices $(4,0),(-4,0)$
    transverse axis 8, conjugate axis 6
    asymptotes $y=\pm3/4x$
    $9x^2-16y^2=144$

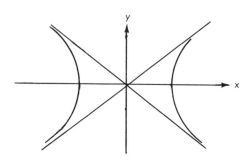

3. $9y^2-16x^2=144$
    Center $(0,0)$, Foci $(0,5),(0,-5)$
    $e=5/4$, Vertices $(0,4),(0,-4)$
    transverse axis 8, conjugate axis 6
    asymptotes $y=\pm3/4x$

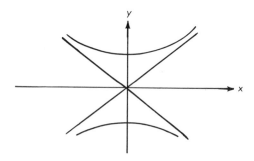

5. $(3x^2)/10-(y^2/5)=1$
    Center $(0,0)$, Foci $(5/\sqrt{3},0),(-5/\sqrt{3},0)$
    $e=\sqrt{10}/2$, Vertices $(\sqrt{10}/3,0),(-\sqrt{10}/3,0)$
    transverse axis $2\sqrt{10}/3$, conjugate axis $2\sqrt{5}$
    asymptotes $y\pm\sqrt{3}/2\ x$

7. Center $(-1,1)$, Foci $(-1+\sqrt{13}),(-1-\sqrt{13})$
    $e=\sqrt{13}/2$, Vertices $(-3,1),(1,1)$
    transverse axis 4, conjugate axis 6
    asymptotes, $y=3/2x+5/2$, $y=-3/2x-\frac{1}{2}$

9. $(x-1)^2-4(y+2)^2=5$
    Center $(1,-2)$, Foci $(7/2,-2),(-3/2,-2)$
    $e=\sqrt{5}/2$, Vertices
    $(1-\sqrt{5},-2),(1+\sqrt{5},-2)$
    transverse axis $2\sqrt{5}$, conjugate axis $\sqrt{5}$
    asymptotes, $y=\frac{1}{2}x-5/2$, $y=-\frac{1}{2}x-3/2$

11. Center $(1,-2)$, Foci $(1,-2+\sqrt{13})$, $(1,-2-\sqrt{13}$
    $e=\sqrt{13}/3$, Vertices $(1,1),(1,-5)$
    transverse axis 6, conjugate axis 4
    asymptotes $y=2/3x-8/3$, $y=-2/3x-4/3$

13. $7x^2-9y^2=63$ 15. $4(x+1)^2-5(y-1)^2=20$ 17. $(y+2)^2-4(x+5)^2=4$

19. $9y^2-4x^2=128$

EXERCISES 7.6

1. ellipse   2. hyperbola   3. hyperbola   5. ellipse   7. hyperbola

8. parabola   11. parabola   13. $r=4/(1+\sin\theta)$ 15. $r=4/(3+\sin\theta)$

REVIEW OF CHAPTER SEVEN

5. $x^2+y^2-4x+6y-3=0$ 6. $C=(-2,3)$, $r=5$ 8. $A_1=(0,4)$, $A_2=(0,-4)$

9. $4x^2+9y^2=36$ 11. $(2,-1)$, $(2,-1+\sqrt{3})$, $(2,-1-\sqrt{3})$; 4

13. 1   15. $(33/32,2)$; $(1,2)$, $x=31/32$ 17. two   19. $9x^2-16y^2=144$

TEST SEVEN

1. $C=(3,-2)$, $r=\sqrt{17}$ 2. $(x-3)^2+(y-4)^2=16$

3. a. 6   b. $(\sqrt{6},0),(-\sqrt{6},0)$ c. $A_1=(3,0),A_2(-3,0)$ d. $\sqrt{6}/3$

4. $16x^2+17y^2=288$ 5. $(x-2)^2=12(y-2)$

6. a. $(6,-5)$ b. $(95/16,-5)$ c. $x=97/16$ 7. $[(y+1)^2/5]-[(x-1)^2/6^2]=1$

8. $r=4/(1\pm2\cos\theta)$ 9. ellipse

APPENDIX A

EXERCISES A.1

1. (b) (c) (d)   2. (a) (d)

3. a. $\{1,2,3,4,5,6,7,8,9,10\}$
   b. $\{Atlantic, Pacific, Indian, Arctic, Antarctic\}$
   c. $\{Carter\}$
   d. $\{m,a,t,h,e,i,c,s\}$

6. a. 2   b. 0   c. ±2 7. $\{1,3,4,5,6,8\}$

9. $\{1,3,5\}$ 11. $\{0,6,7,8,9\}$ 13. $\{0,2,4,6,7,8,9\}$

15. $\{1,3,5\}$ 17. U   20. U   21. B

23. $A=\{a,b,c\}$, $B=\{a,c,d\}$, $C=\{a,b\}$, $D=\{1\}$; C and D are not unique.

EXERCISES A.2

1. $\log_2 16 = 4$

2. $\log_3 243 = 5$

3. $\log_5(1/25) = -2$

4. $\log_{1/2} 8 = -3$

5. $\log_{10} 0.0001 = -4$ 6. $\log_{16} 4 = \frac{1}{2}$

7. $\log_{32} 8 = 3/5$

8. $\log_5 1 = 0$

9. $3^4 = 81$

10. $10^3 = 1000$

11. $4^{-1/2} = \frac{1}{2}$

12. $10^{-2} = 1/100$

13. $(\frac{1}{3})^{-2} = 9$

14. $16^{-3/4} = 1/8$

15. $-\frac{1}{3}$

16. 2

17. $\frac{1}{2}$

18. $\frac{1}{3}$

19. 4

20. 1/8

21. 9/2

22. 17/7

23. 5

24. 10/7

25. 9/2

31. 0

33. $\log_3 4$

34. 0

35. a. -0.176 b. 0.631   c. 0.903   d. 0.0272   e. 1.176   f. 1.477

37. 2.6294

39. 1.1987

41. $0.1303 - 1$

43. 702

45. 0.451

49. 3.5

51. 0.0245

55. $0.9710 - 4$

57. 352.6

61. 0.007741

63. 36.70

65. 0.04858

69. 104,600

75. 0.04260

77. -0.0433

79. 2.085

81. 0.2524

83. 8.966

INDEX